43.99

ELECTRONICS WITH DISCRETE COMPONENTS

ELECTRONICS WITH DISCRETE COMPONENTS

Enrique J. Galvez
Department of Physics and Astronomy
Colgate University

WILEY

John Wiley & Sons, Inc.

Vice President and Executive Publisher	Kaye Pace
Executive Editor	Stuart Johnson
Marketing Manager	Christine Kushner
Senior Production Manager	Janis Soo
Associate Production Manager	Joyce Poh
Assistant Production Editor	Yee Lyn Song
Cover Designer	Seng Ping Ngieng
Cover Photo Credit	Enrique J. Galvez

About the Cover: The background is a photo of a multicrystalline solar cell. Thin lines are electrodes and the different shades are the single-crystal sections of silicon that grow randomly in the manufacturing process.

This book was set in 10/12 Times Roman by MPS Limited and printed and bound by Courier Westford. The cover was printed by Courier Westford.

This book is printed on acid-free paper. ∞

Founded in 1807, John Wiley & Sons, Inc. has been a valued source of knowledge and understanding for more than 200 years, helping people around the world meet their needs and fulfill their aspirations. Our company is built on a foundation of principles that include responsibility to the communities we serve and where we live and work. In 2008, we launched a Corporate Citizenship Initiative, a global effort to address the environmental, social, economic, and ethical challenges we face in our business. Among the issues we are addressing are carbon impact, paper specifications and procurement, ethical conduct within our business and among our vendors, and community and charitable support. For more information, please visit our website: www.wiley.com/go/citizenship.

Evaluation copies are provided to qualified academics and professionals for review purposes only, for use in their courses during the next academic year. These copies are licensed and may not be sold or transferred to a third party. Upon completion of the review period, please return the evaluation copy to Wiley. Return instructions and a free of charge return mailing label are available at www.wiley.com/go/returnlabel. If you have chosen to adopt this textbook for use in your course, please accept this book as your complimentary desk copy. Outside of the United States, please contact your local sales representative.

Library of Congress Cataloging-in-Publication Data

Galvez, Enrique Jose, 1956-
    Electronics with discrete components / Enrique J. Galvez.
        pages cm
    Includes index.
    ISBN 978-0-470-88968-8
        1. Electronics—Textbooks.   2. Digital electronics—Textbooks.   3. Analog electronic systems—Textbooks.
    I. Title.
    TK7816.G35   2013
    621.3815—dc23

                                                                                    2012003004

Printed in the United States of America

10 9 8 7 6 5 4 3 2 1

I dedicate this book to Mary, Ricky, Samantha, Daniel, and Elena

PREFACE

This text is designed for a one-semester course on electronics. Its primary audience is second-year physics students, but it can include students from other disciplines or levels who understand elementary notions of circuits and complex numbers. Most physics programs, especially those in liberal arts colleges, can afford only a one-semester course in electronics. Electronics is a vital part of a curriculum because it trains students in a basic skill of experimentation. With this knowledge, students can design circuits to manipulate electronic signals or drive mechanical devices. An electronics course also gives students a basic understanding of the inner workings of electronics instruments. Thus, an electronics course prepares students for advanced laboratories and, ultimately, experimental research.

Because of the nature of the topic, the course must have a huge hands-on component. Electronics is learned by experience. At Colgate University, we have been teaching a course that meets two days a week, with a one-hour lecture followed by a two- to three-hour lab. In the lab, students build circuits that closely follow the topic of the class. We have put special effort into making those labs instructive but at the same time interesting, empowering, and fun. We made a special effort to introduce transducers in the labs, highlighting applications. Today's students live around black boxes, mostly ignorant of the circuits that lie within them. Our recent experience tells us that students find the discovery of how those boxes work, or even the task of building them, extremely interesting, rewarding, and useful. Thus, we can use this "revelation" as a way to motivate students to learn electronics.

Instructors who adopt this text may have labs in place and may not have use for the labs in this book. However, the experiments listed may give instructors ideas to renew or modify the labs in place. In addition to the normal curricular plan, we devote two weeks in the middle of the semester and two weeks at the end of the semester to unscripted projects, in which students design the device of their choice. This is where students learn tremendously and enjoy the experience. Their ambition to build the device of their choice pushes them to invest much energy and time, and along the way, they learn invaluable aspects of building devices, such as creating new designs and troubleshooting. In the first project, students do mostly digital work (more on this choice below), but they still use a little bit of analog, because they need switches or pushbuttons for digital inputs and light-emitting diodes (LED) for digital outputs. In the second project, students do mostly analog work, but they can combine analog and digital electronics. Whatever the case, students end up doing amazing projects. Some of the analog projects can be combined with real computers, but this is an aspect that we do not cover here. If lab PCs have interface cards, the projects will be more powerful. A word of caution from experience: Make sure that the project does not become a computer project. Although knowing programming is not that bad of a goal these days, it is not the objective of this course.

The text is divided into two parts: digital and analog. In each part, we cover the essential components needed to understand and design circuits with discrete components. We cover the digital part first. This may seem like heresy to some instructors, but I urge them to reconsider the concept. Covering digital first makes sense because digital electronics focuses mostly on logic. The topic is not as intellectually demanding as analog. Besides a few rules of thumb for wiring, students have little need to know about the currents that flow through the gates or even the analog circuits that make up those gates. Later in the semester, after covering the analog part, the class revisits the details of gates. The digital part is demanding on wiring practices, but not on conceptual understanding. This way, students get early exposure to demanding circuits and are forced to embrace systematic wiring practices. By the time students reach analog, they no longer have trouble wiring and powering circuits. It makes sense to cover analog after digital because students end with the understanding of the complexity and importance of analog. Otherwise, students would get the wrong message: Since analog is not needed to do digital, it is unnecessary altogether. An instructor who strongly disagrees with this strategy could swap the two parts without major logistic complications, but he or she would have to continue to emphasize analog concepts throughout the digital part.

The content of this text borrows ideas on the organization of topics from two classics in the field: *Digital Design*, by M. Morris Mano, and *The Art of Electronics*, by Paul Horowitz and Winfield Hill. The chapters are designed so that they take an integral number of days. Labs may also extend one day, and in digital, several labs build upon the circuit of the previous lab. The topics of the specific chapters go as follows. The first chapter, "The Basics," reviews the fundamentals of electricity and electrical components. It brings the student, especially the nonphysics major, up to speed with the physics and basics of electric circuits. The second chapter, "Introduction to Digital Electronics," covers digital signals and electronic gates. It is followed by two chapters on combinational logic, namely "Combinational Logic" and "Advanced Combinational Devices." They are followed by a chapter titled "Sequential Logic," which emphasizes counting circuits, and an important application in memory. Throughout this part, we include tables of integrated circuits that are useful for designing circuits. A rack of ICs of various types is vital in an electronics lab. The lab exercises use a "logic board," which is a homemade or commercial box with switches that generate input states, and LEDs to display output states. Appendix A gives the details of this device and its construction. Some versions of these boards are commercially available. If time permits, the instructor may consider other adventures, such as microcontrollers and interfacing using Labview, but such endeavors are specialized to particular equipment for which there is no uniform agreement. Instead of attempting a partial or incomplete description, we do not cover those at all.

The analog part starts with the chapter "AC Signals." It covers a more sophisticated analysis of circuits than the first chapter and centers on the use of complex numbers for defining signals and impedances. We find this advantageous and practical. To complement this, we include a short introduction to complex numbers. It ends with an important concept to students: Thevenin equivalent circuits. Throughout, this part reduces circuits to single-loop modules, building up the concepts of input and output impedance. We follow with the chapter "Filters and the Frequency Domain," where the role of frequency and frequency response comes to the surface. The use of multiple filter stages underscores the role of source and load impedance. At the end of this chapter, we insert a section on Fourier Series. This is important because electronics' processing of signals can be understood easily at the single frequency level. Therefore, knowing the decomposition of a complex signal into its frequency spectrum is vital in understanding the frequency response of a circuit. This part can be skipped if the curriculum already contains Fourier series. The chapter that follows, "Diodes," starts with a physical explanation of semiconductors that gives the student an intuitive and informed basic understanding of the physics of these materials. It emphasizes nonlinear responses and the use of the load line, and ends with an application on the design of power supplies, among other diode tricks. The chapter titled "Transistors," covers both bipolar-junction and field-effect transistors. Because operational amplifiers are much better suited for signal conditioning, we do not cover in detail some of the traditional circuits on biasing the transistor. Increasingly, modern devices use field-effect transistors instead of bipolar transistors, so we give both nearly equal coverage and focus on power drivers, followers, and current sources. These are applications that even operational amplifiers cannot deliver and in which transistors have rightful place. The final part of analog is the experimenters delight: "Operational Amplifiers." We give ample coverage to numerous circuits, plus we use them to smuggle in other interesting topics, such as comparators and feedback. We wrap up with a chapter that interfaces digital and analog signals and transducers, in "Connecting Digital to Analog and to the World."

At the end of most chapters is a section titled "Lab Projects" that contains many interesting circuits that have been proven to work well for instruction. Many of them have interesting twists that make the experience a fun one. I like to follow this motto: "Let the kids have fun." If they do, they will learn electronics. Our tests also have a practical component. When students work in groups there is a danger that they are passive and let their partner(s) do valuable laboratory know-how. To force them to be active participants, we test them individually on building simple circuits. The final section of each chapter is titled "Practicum Test." It gives questions that we have often asked on simple aspects of the lab that students should know. This includes powering components and diagnosing signals with the oscilloscope. The goal is for each student be able to do every task and not leave any activity, and know-how, to his or her partner.

I owe immense gratitude to Joseph Amato. Together we designed this course almost 20 years ago. His prolific expertise and creativity led to the design of a number of lab experiences described in this text. I also want to thank Wes Walters for selling me the idea of covering digital before analog; Dave Glenar, Ken Segall, and Steve Slivan for ideas for labs and problems; Juan Burciaga and Danielle Solomon for useful suggestions and edits; Timothy Kidd, M.K. Kim, Bryan Suits, Christos Velissaris and other anonymous reviewers of the drafts of this book for their valuable advice; and Samantha and Daniel Galvez for helping Dad with aspects of this project.

E.J. "Kiko" Galvez

CONTENTS

CHAPTER 1

THE BASICS

Contents

1.1 FOREWORD: WELCOME TO ELECTRONICS!

This book is primarily geared for physics students, but nonphysics students with some basic physics and math can understand it. Our focus is not physics. We cover the fundamental physics to provide a foundation, but our primary concern is the electronic devices. The good news is that you will learn how some of those black boxes with glitzy lights work and, going beyond that, how to build some of your own boxes. You will discover that the most complicated machines—computers—are as logical as the gears in a bike. Often in this book, we do not approach the subjects as precisely as physicists treat other subjects. For example, using 10 percent accuracies or even factors of 2 for device parameters is usually fine in electronics. Electronics also involves a lot of details, so do not get overwhelmed. Experience will help you distinguish the important details from the less important ones, but still be prepared to take in a lot!

Electronics should be learned from the ground up. Although one can easily go a long way in electronics by knowing some fundamental concepts and understanding how the devices work, a solid foundation in electricity and magnetism is important for an in-depth understanding. The goal of this chapter is to cover the underlying physics, in case the reader lacks a previous foundation in electricity and magnetism. Because electronics is closer to engineering than physics, we are interested less in learning the underlying physical laws as ends in themselves, and more on understanding devices and how they work. Take this course also as an opportunity to learn that every device is based on important physical principles. Knowing those principles will give you an increasing edge in mastering electronics.

We start with fundamental concepts and work our way to devices. As we gain some speed, we will move into elementary circuits.

1.2 CHARGE AND POTENTIAL

Electric charge is a fundamental property of matter that is responsible for most of its structure as we know it. Taking central stage in our electronic world is nature's premier fundamental particle: $\eta\lambda\epsilon\kappa\tau\rho\upsilon$. If you have taken enough physics, you will read *electron*. It is the Greek's name for amber, as the ancient Greeks recognized the curious (electrical) properties of amber. Not only do we take for granted the existence of electrons, which are in everything we see and touch (ourselves included), but we "feel" their presence directly with the jolt of static electricity that we get on a dry day. Electrons are simple: They have a mass, and, for the most part, they behave as point particles. Despite trying to find a dimension to them, we have been unsuccessful. Electrons do not always behave like particles: Sometimes they behave as waves. When they do so, people studying them have to figure out not only what the electrons are doing, but what they really are.

Electrons' most important property is their charge. For some fateful reason that originated with the cleverness and wit of Benjamin Franklin, the charge of the electron is labeled as negative. The electron has one unit of elementary charge, which is $q_e = -e$, where $e = 1.6 \times 10^{-19}$ C, with C being the SI unit of the electrical charge, the Coulomb. This value is quite precise and is deemed fundamental by physicists. Do not bother trying to discern the meaning of "fundamental"—it is a physicist's way of saying, "It is what it is and we do not know why." A beautiful story of experimentation involves the measurement of the electronic charge by Robert Millikan. Electrons also have spin, which is at the root of many interesting phenomena, such as magnetism. However, for all purposes that concern us, electrons are simple and have a definite charge.

Atoms have a nucleus that has a charge of the opposite sign: positive. The nucleus is formed by two particles: protons and neutrons. The only exception is the most abundant isotope in the universe, hydrogen, which has only one proton as a nucleus. Protons have a charge, $q_p = +e$, and neutrons have no charge $q_n = 0$. Note that the magnitude of the charge of the protons is *exactly* the same as that of the electron; nature as we know it would not exist if the charges of electrons and protons did not have the exact same magnitude. The properties of matter rely on the exact electrical neutrality of atoms. Protons and neutrons are made of *quarks*, each of which has a fractional charge: The *up* quark has a charge $q_u = +(\frac{2}{3})e$ and the *down* quark has a charge $q_d = -(\frac{1}{3})e$. This way, a proton is made of two ups and one down, and the neutron consists of one up and two downs. Yet for all the fascinating consequences of the existence of quarks, we never see them by themselves because of a strong attractive force that increases with distance. So for all practical purposes, protons and neutrons are whole particles.

Atoms are neutral, but the electrons buzzing around the nucleus follow special rules of behavior dictated by quantum mechanics. We say "buzzing around" because we know they go around the nucleus, but we do not know exactly how. We cannot find out in a deterministic classical way how they move (such as describing nice ellipses). Instead we can only know where they are likely to be, probabilistically; for all we know they can move around in whichever way they please. However, one thing is certain: electrons buzz around always experiencing an attractive force with the positive nucleus and a repulsive force with fellow electrons. Within the nucleus protons still repel each other due to electric forces, but at the short distances within nuclei they are attracted to each other by the stronger nuclear force.

The nucleus is small—100,000 times smaller than the outlying orbits of the electrons in atoms. Atoms are symmetrically neutral when left alone, but when they are pushed against each other the electrons rearrange and atoms are no longer neutrally symmetric: The sides facing each other are more positive on average, and the sides facing away are more negative. A strong repulsion ensues, preventing atoms from getting too close to each other. As a result, matter is mostly made of empty space: Atoms are held away from each other at distances comparable to the sizes of the outer orbits of the electrons, which are point particles, with a tiny nucleus located somewhere inside. This is why neutrons can go a long way through matter without stopping. Electrons' strong interaction with light, an electromagnetic disturbance, makes matter mostly opaque to light (with noted exceptions, such as glass), but electrical forces make matter seem solid when in actuality it is not.

Another property of atoms is that electrons can leave their home atoms to join foreign atoms and make ions (atoms with a net charge). When we rub a plastic (such as a comb or pen) with our sweater on a dry day, we end up with a negatively charged plastic and a positively charged sweater. Electrons from the sweater jump to the plastic when we rub the two together. By applying clever techniques, we can use this effect to charge objects deliberately. Other charged objects in the vicinity then experience forces and react accordingly. Although the concept of force is a useful one to conceptually understand what is happening, it is not convenient for a quantifying the events. It is more practical to use energy arguments: If two objects have the same charge, then as they get close to each other their electrical potential energy increases. If we let them go, they will repel each other, converting the potential energy into kinetic energy and going to places where the potential energy is lower (of course, energy must be conserved).

If we have a charged object in a fixed position and place another charge in its vicinity, then the latter will have a positive potential energy and experience repulsion if it has the same sign as the charge of the fixed object; if it has a charge of the opposite sign, or negative potential energy, it will experience attraction. Thus, the potential energy depends on the charge of the two objects. To separate cause from effect, we define the concept of *electric potential*, or *voltage*. Electric potential is the electrical potential energy per unit charge. To get the potential energy of an object with a charge q at a point with potential V, we use this relationship:

$$U = qV \tag{1.1}$$

The units of potential are *volts* (V).

A battery is a device that uses a chemical process to create a separation of charges that maintains a fixed potential. When it is connected to a circuit, it provides as many charges as needed to maintain that voltage. Different batteries differ in their capacity to replenish those charges. Cylindrical household batteries produce a potential of 1.5 V. Their ends are clearly marked: The end with the bump is positive and the flat end is negative. They differ in their ability to source charge, which goes in the order AAA, AA, C, and D. Traditional one-time-use batteries are made of zinc-carbon (cheap) or alkaline (higher quality). Other types of batteries produce other voltages. For example, flat, round lithium-iodide batteries (CR 2016) output 3 V. This is only an incomplete list because of the huge variety of technologies and battery outputs. Batteries play an important role in our lives today with the widespread use of portable devices such as cell phones, iPods, and laptops. One-time-use batteries lose their charge as the chemical reactions in them cause irreversible changes in the battery. Once the charge-producing elements are transformed, they are no longer useful. Rechargeable batteries, such as nickel-cadmium, nickel-metal hydride, and lithium-ion batteries are increasingly replacing one-time-use batteries. Conversely, the lead-acid batteries of cars have existed for a long time. In the recharging process, charges are forced in the direction opposite to the direction they take when the battery is operating normally. In doing so, they produce inverse chemical reactions that revert the chemical elements of the battery to their original amounts. Some voltage supplies rely on the electric utilities (for example, wall outlets). They use a different technology that we will examine in detail later when we cover power rectification (Chapter 8—Diodes) and regulation (Chapter 10—Operational Amplifiers).

We need to cover one more concept in the physics of electricity: electric fields. This mathematical concept assigns a vector to all points in space whenever charges are present somewhere. The electric field E is the force-per-unit charge that a charged object will experience at any given point. We can relate it to the potential the following way: Its magnitude is the rate of change of the potential, and its direction is the direction in which the potential is decreasing most rapidly (mathematically, the field is the gradient of the potential). In one dimension, then, $E = -\frac{dV}{dx}$ (the minus sign indicates that the electric field points in the direction where the potential is *decreasing*). The unit of electric field is V/m. The force is the charge times the field: $F = qE$. Thus, if the potential in a region of space varies from point to point, then that means that if we put a charge in that region, it will experience a force. The strength and direction of the field are determined by how potential varies with the spatial coordinates.

1.3 CAPACITORS

The first electrical component that we consider is the *capacitor*. The simplest capacitor consists of two parallel metal plates of area A that are separated by a distance d. We charge the capacitor by moving the charges from one plate to the other. As soon as each plate has charges of opposite sign (see Figure 1.1), the plates will have a potential difference between them,

$$V = \frac{q}{C} \tag{1.2}$$

Here C is the *capacitance*, which depends on the geometry and composition of the device. The units of capacitance are farads (F). This equation is normally derived in introductory electricity and magnetism textbooks, so we do not repeat that here; suffice it to say that the derivation involves building the potential energy by calculating the work required to move the charges from one plate to the other.

Figure 1.1.   A parallel-plate capacitor with a plate separation d and a charge q.

For the simple parallel-plate capacitor of our discussion

$$C = \frac{\epsilon_0 A}{d} \tag{1.3}$$

where $\epsilon_0 = 8.85 \times 10^{-12}$ F/m is the permeability of vacuum. Most capacitors are a variation of this simple design, so to increase the capacitance, we need to either increase the area of the electrodes (plates) or decrease the distance between them. To get an idea of capacitance values, consider two square metal plates 2.54 cm (1 inch) on the side separated by 1 mm. The capacitance is 5.7 pF. Note that this is a very small value. If we set up the potential between the plates to be 1.5 V, then the plates will have a charge of 8.6 pC .

The jolts that we get on a dry day involve quick transfers of charge through a spark. Air breaks down in an electric field of 3 MV/m. The field must be strong enough to rip an electron from an air molecule. For us to see a spark across our capacitor, we would need a potential of 3,000 volts, which would mean putting a charge of 17 nC on its plates.

We can increase the capacitance of a capacitor by placing a dielectric of constant κ between the electrodes. The capacitance with the dielectric is κC_0, where C_0 is the capacitance with vacuum between the plates. Typical values of κ are greater than 1 and can go as high as about 20 (for Teflon, it is 2.1, and for mylar 3.1). The dielectric also helps increase the potential at which the capacitor breaks down (for Teflon, it is about 60 MV/m, and for mylar, it is about 170 MV/m).

The values of capacitors in a typical circuit fall in the range between pF and μF. Prepackaged capacitors in this range differ in the way the two conductors and dielectric are packaged. In film capacitors, the conductors and dielectric are rolled around each other; in mica capacitors, flat conductor sheets are separated by mica sheets; and in disk capacitors, a ceramic disk is placed between two conductor disks. Capacitances above 1 μF need a significantly smaller value of d to be of a practical size. Electrolytic capacitors and tantalum capacitors have a conductor surrounded by a conducting electrolyte. An oxide layer coating the electrode serves as the dielectric. The small thickness of the insulating layer, as low as a few nanometers and as high as a few micrometers, allows these capacitors to achieve high capacitance values. However, in electrolytic capacitors one conductor *has to be positive relative to the other*. Be sure to properly bias these capacitors when you use them. Otherwise, the dielectric film gets decomposed by electrolysis, and when the thickness reaches zero, you get a short (which can make a big bang, and cause a fire). Figure 1.2 shows the symbol for this type of capacitor.

Capacitors have wide uses in electronics. You will see many applications in this course, ranging from their frequency-dependent response to AC signals to more specialized

Figure 1.2. Symbol for the electrolytic capacitor (left), where one side must be more positive than the other side. The negative plate is drawn curved. On the right is a photo of an electrolytic capacitor in its characteristic cylindrical can shape. Notice that the positive end is clearly marked. Other styles have the negative end marked.

ones, such as rectification and charge storage. Modern memories store digital data (1's and 0's) by charging and discharging micron-size capacitors inside the integrated circuits.

As mentioned earlier, a capacitance of 1 F is very large. Normally, you will use capacitances that are much lower in value. However, beware of the dangers of large capacitors that are charged to a high voltage. High-voltage capacitors, like those inside television tubes and discharge-lamp supplies such as gas lasers, can discharge a deadly amount of charge. Warning signs that explicitly state high voltage normally refer to circuits that contain capacitors charged to dangerous levels.

At the other end, there is no lower limit for capacitance. If your circuit is sensitive to capacitances less than 1 pF, you are in trouble! Capacitances between adjacent wires in a circuit are of this order of magnitude. Ambient vibrations may change the separation between dangling wires in the circuit and create vibration-induced electrical noise, also known as microphonics. The effect appears because the capacitance of the dangling wires changes with vibrations, and the change in capacitance causes charges to flow back and forth to keep the wires charged to the correct amount. Paired conductors and coaxial cables (such as the one used for cable TV) have capacitances of the order of a few to tens of picofarads per foot.

When a capacitor is connected to a voltage source, its plates acquire a charge $q = CV$ ($+q$ in one plate and $-q$ in the other plate). Thus, a larger capacitance gives a greater capacity to store charge. When two capacitors are connected in *parallel*, with their two ends joined together, you can store more charge.

1.4   ELECTRICAL CURRENT

The electrical current is defined as the flow of charge—or, more specifically, the amount of charge that crosses some surface per unit time. It is represented by

$$I = \frac{dq}{dt} \tag{1.4}$$

where q is the charge and t is time. The unit of electric current is the ampere (A), which is the same as C/s. To describe electrical currents in bulk matter, it is convenient to define the *current density*, which is the charge flow per unit area: the amount of charge that crosses a transverse unit area per unit time. Let us first consider an idealized situation: a river of particles with charge q all moving in one direction with velocity v. If there are n charges per unit volume, then the current density is:

$$J = nqv \tag{1.5}$$

If the river of particles crosses a transverse surface of area A, then the total current crossing the surface is:

$$I = JA \qquad\qquad (1.6)$$

Because the charges can be positive or negative, we define the *direction of the current* to be the *direction in which positive charges move*. That is, if the charges are positive, then the current goes in the same direction as the velocity of the particles. However, if the charges are negative, then the current goes in the direction *opposite* to the velocity of the charges. An example of electrical current is an electron tube, such as in an old TV. Although these are increasingly becoming obsolete, they have been of huge importance in our history. The "tube" generates electron beams in a sealed container that are sent from the back of the tube to the front, the screen, where electrons excite phosphor atoms as they crash into them. The electrons in the tube TV constitute an electric current whose direction is opposite to the direction in which the electrons are traveling. In metals, to be discussed in the next section, electrons are also the flowing particles, so the currents in wires are also flowing in a direction opposite to the direction of the actual flow of electrons. Semiconductors have *holes*, which are "vacancies," or the lack of electrons. These holes behave like positive charges, so they move in the same direction as the current.

Interestingly, when an electronics-literate person looks at a circuit diagram with electron tubes, semiconductors, or other devices, he or she follows the electric current, not the particles. Although we are asking you to understand the fundamentals, we are also asking you to *ignore* some of the details when you are looking at a circuit in a general sense, such as what the charged particles are actually doing. This keeps you from getting distracted by too much unnecessary information. Pragmatism goes a long way in electronics.

The magnitude of the currents flowing through most of the circuits in the labs such as the one for this course are of the order of tens of milliamperes. Currents above a few hundred milliamperes require high-power components. Table 1.1 gives a few examples of situations involving currents of different magnitudes. Currents of the order of amperes must be taken more seriously, with special design and safety considerations.

Table 1.1.   Electrical Currents in Different Situations

Value	Situation
10 nA	Induced current in a radio antenna
1 μA	Current generated by a photodiode
1 mA	Current flowing through a TTL gate
50 mA	Lethal dose of current flowing through the heart
200 mA	Current in an electron tube
1 A	Current flowing through an incandescent light bulb
15 A	Current flowing through the heater of a hair dryer

1.5 RESISTORS

Metals and metal alloys have intrinsic properties that allow easy conduction of electricity. They lie prominently in a class of materials that we call conductors. The atoms in a metal are bonded in what it is called metallic bonding. In this type of bonding, all the metal atoms in the material share some of the electrons, so these electrons are free to move within the material. However, do not think that electrons in a metal are moving smoothly from one end of the metal to the other. Quite the opposite is true: Their trajectories are erratic as they bounce off everything in their way inside the metal, such as atoms, defects, and other electrons. The distance that electrons travel between bounces is actually very short. For example, for copper at room temperature, this distance is of the order of tens of nanometers. Despite these impediments, metals can easily conduct electricity, and are the most common materials used for wires.[1]

When we connect the ends of a wire to a source of constant potential, such as a battery, something remarkable happens. The charges rearrange quickly through the metal, creating a longitudinal electric field along the wire, regardless of how windy it is. As a consequence, the electric potential distributes uniformly along the wire (provided that the wire is uniform). This electric field along the length of the wire gets the electrons to respond accordingly. However, they do not move as you may think (along the wire). The electrons bounce, still randomly, but now with a net overall motion along the wire. Although electrons travel very fast—close to the speed of light between bounces—the net speed along the wire, also known as the *drift velocity*, is small. For example, for copper in typical conditions this speed is of the order of millimeters per second. Why, then, when we flip the switch on the wall do the lights come on seemingly instantaneously? The answer is that *all* the free electrons in the circuit are doing this. It is like opening the valve of a garden hose when the hose is full of water: The water starts coming out of the hose almost immediately because the source of water pushes the water that is already in the hose.

The electronic properties of materials have other classifications. Insulators, or dielectrics, have no free electrons to support a current. These, of course, play an important role in circuits by not allowing currents to flow in unwanted paths and in insulating conductors from each other. A wide range of materials are insulators, such as glass and plastics. In semiconductors the conductivity can be controlled by adding impurity atoms during the fabrication process. This allows for the production of very important devices, such as diodes and transistors. We will defer a discussion of semiconductors to Chapter 8 (Diodes).

As mentioned earlier, when an electric field is established in a conductor, a current ensues. This current can be expressed by Equation 1.5, where v is now the drift velocity. However, this equation is "sort of" correct because the net motion is along the wire, but it is not how the original equation was derived. In any case, it is better to use a different form of that equation. Because the drift velocity depends on the electric field and electronic properties of the material, we can rewrite Equation 1.5 the following way:

$$J = \sigma E \qquad (1.7)$$

This equation is important because it groups the properties of the material into σ, which is called the electrical conductivity, and separates it from the external stimulus, the electric field E. This equation is a fundamental version of *Ohm's law*. The *resistivity* of a material is the opposite of the conductivity because it measures the material's opposition to a current. It is given by this equation:

$$\rho = \frac{1}{\sigma} \qquad (1.8)$$

---

[1]A fascinating type of conductor, called *superconductor*, creates a situation in which electrons travel through the material without opposition. We do not discuss these because they are generally impractical to use in most electronics devices—they have to be cooled to about 4 K. Superconductors are used in specialized applications that require high currents and no dissipation, such as in the magnets used in magnetic resonance imaging.

We can rework Equation 1.7 in an interesting way: If we express the magnitude of the electric field in terms of the voltage across the conductor with $E = \frac{V}{d}$, where d is the length of the conductor, and the total current by Equation 1.6, then we can leave it as an exercise to derive the following relationship:

$$I = \frac{A}{\rho d} V$$

or

$$I = \frac{V}{R} \qquad (1.9)$$

where

$$R = \frac{\rho d}{A} \qquad (1.10)$$

is the *resistance* of the particular section of conductor we are considering. Equation 1.9 is the more well-known version of Ohm's law. The unit of resistance is the ohm (Ω). Copper and silver have the lowest resistivities of metals, at 1.7×10^{-8} Ωm and 1.6×10^{-8} Ωm, respectively. Resistivities quickly increase for other metals: For aluminum, it is 2.7×10^{-8} Ωm, and for iron, it is 9.7×10^{-8} Ωm. Thus, for all practical applications, copper is the material of choice for conductor wires.

Conductors then present a small but finite opposition to current flow. For example, the resistance of 24-gauge wire (0.511 mm diameter) used for making circuits in a breadboard is 84.2 mΩ/m. Because the resistance depends on the wire's cross-section, the thicker 22-gauge wire (0.644 mm diameter) also common in the lab has a resistance of 53 mΩ/m; the ratio of the resistances (1.59) goes with the inverse ratio of the areas, in accordance with Equation 1.10. If we connect two electrical components in a circuit with a 24-gauge copper wire that is 5 cm long, then the resistance of that section of cable is 4.2 mΩ. Although finite, this resistance is much smaller than the resistances of other circuit components, which are typically of the order of thousands of ohms. So for all practical purposes, we do not account for conductors' resistance in figuring out circuits.

Poor conductors, called resistors, are useful because they allow us to control the amount of current flowing through a circuit. Metals, carbon, and other semiconductors are commonly used as resistors. In metals, the abundance of impurities and defects greatly diminishes the conductivity, so they can be used as resistors. In semiconductors such as carbon, the mechanisms of conduction are more complex than for metals. Not many electrons are readily free for conduction, as in metals, leading to higher resistivities. In metals, higher temperatures mean more collisions and higher resistance. However, in semiconductors, as the temperature increases, more electrons are freed and resistance actually decreases.

The symbol for the resistor is the most recognizable feature of a circuit diagram: a squiggle line. The voltage across a resistor is sometimes called the *voltage drop*. Figure 1.3 shows its polarity. The upstream end is positive relative to the downstream end.

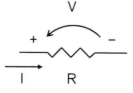

Figure 1.3.   Symbol for the resistor, with the label of the potential drop for the given current direction.

You can also think that one part is more positive than the other one in order to establish an electric field along the resistor. The higher resistance of a resistor compared to conductors implies that, in a resistor, electrons go through a greater number of collisions within the material.

A resistor's value is labeled by a set of color bands, shown in Figure 1.4. If you are making circuits in the lab, you must learn this labeling. Normally there are four colored bands. The first three determine the value of the resistance and the fourth one determines the accuracy of that value, or the tolerance.

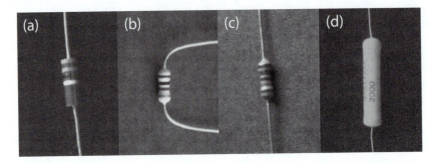

Figure 1.4.   Photo of resistors of various types: (a) carbon composite, (b) carbon film, (c) metal film, and (d) wire-wound.

The value of the resistor is given by

$$R = AB \times 10^{C} \tag{1.11}$$

where AB is a two-digit number. The values of A, B, and C are given by the colors of the first (from the edge), second, and third bands, respectively. The color-to-number correspondence follows the rainbow, as shown in Table 1.2. The fourth band, represents the tolerance: white through brown in Table 1.3. For example, the color bands of the resistor in Figure 1.4(b), red, green, brown, and gold, represent a resistance of $250\,\Omega \pm 5\%$. Resistors with higher precision require more significant digits, so they have five bands (see Figure 1.4(c)),

Table 1.2.   Color Coding of Resistor Values

Value	Band Color
0	Black
1	Brown
2	Red
3	Orange
4	Yellow
5	Green
6	Blue
7	Purple
8	Gray
9	White

three for the mantissa, one for the multiplier, and another one for the precision. The fifth-band tolerance is given by colors green through gray in Table 1.3.

Table 1.3.   Color Coding of Tolerances of Resistor Values

Value	Band color
20%	No band/white
10%	Silver
5%	Gold
2%	Red
1%	Brown
0.5%	Green
0.25%	Blue
0.1%	Purple
0.05%	Gray

Resistors can be of several types. The most common and most inexpensive resistors are made of carbon. The oldest types of resistors are carbon composition. These are made of graphite in bulk but are no longer popular, except in high-power applications. The most popular resistors today are carbon-film resistors because they are inexpensive to make and work well enough for most common applications. You can tell carbon-composite from carbon-film resistors because of their shape: Carbon-composite resistors are straight cylinders, as in Figure 1.4(a), whereas carbon-film resistors somewhat bulge at the ends, as shown in Figure 1.4(b). If the application requires low electrical noise and precise resistor values, metal-film resistors are the better option. Besides high tolerance and less electrical noise, they have good temperature stability. Carbon has poor temperature stability because the number of electrons available for conduction varies with temperature. Another type of film resistor is metal oxide, which tolerates higher powers. One recognizes metal-oxide resistors because they are green, as opposed to carbon ones, which are brown. High-current applications also have wire-wound resistors that use a resistive conductor wound around a cylindrical structure. The latter are not recommended for high-frequency applications because of their inductive properties (see the next section). They are conspicuously bulky in appearance and have a ceramic body with no color bands, as shown in Figure 1.4(d). High-wattage resistors of this kind have a hollow interior for heat dissipation. Nichrome wire is a popular high-resistance wire made of a nickel-chromium-iron alloy that is used in heating applications. For example, 24-gauge (0.5-mm) wire has a resistance of about 1.6 Ω/ft. Passing a current of 2 A heats the wire to about 200° C (you can find table listings of these values on a number of commercial web sites). A fun application of nichrome is the "hot wire," in which a piece of wire is clamped to a U-shape wooden piece and its ends are connected to a low-voltage high-current source so that a current of several amperes passes through it. The hot wire can then be used to cut Styrofoam easily and make creative sculptures.

Other interesting types of resistors are *photoresistors* and *thermistors*. Both are made of semiconductors, which, as mentioned earlier, have a limited number of electrons free for conduction. However, these have another group of electrons that can become available for conduction. You will see the physics of this later. The point is that, in both cases, an external agent, light for photo resistors and temperature for thermistors, frees electrons for conduction, so the greater the external stimulus, via either light or thermal motion, the smaller the resistance. We can use these resistors as transducers to measure illumination and temperature.

1.6 MAGNETIC DEVICES

Magnetic fields and magnetism have important applications in electronics. In this section we describe the fundamentals and some applications.

1.6.1 Magnetic Fields and Coils

Electrical currents produce magnetic fields. The simplest case is that of a straight wire carrying a current I. The magnitude of the magnetic field produced by the current in the wire is

$$B = \frac{\mu_0 I}{2\pi r} \tag{1.12}$$

The magnetic field is a vector. Its direction is tangential to a circle of radius r centered about the wire and lying on a plane perpendicular to the wire. The sense of this circulation of magnetic field is as shown in Figure 1.5(a), and it follows the right-hand rule: The thumb points in the direction of the current, and B circulates in the same sense as the wrapping hand. The unit of magnetic field is the tesla (T), and $\mu_0 = 4\pi \times 10^{-7}$ Tm/A is a constant called the permeability of vacuum.

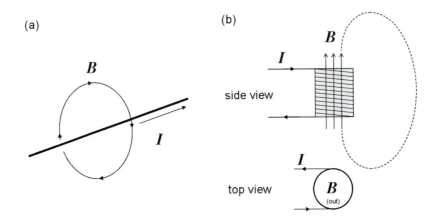

Figure 1.5.   Illustrations of the magnetic field produced by current-carrying wires: (a) straight wire, and (b) coiled wire (solenoid).

The magnetic field that wires produce can be concentrated by wrapping the wire into a coil, such as the one in the figure. The magnetic field inside the coil bundles and has a magnitude

$$B = \mu_0 n I \tag{1.13}$$

where n is the number of turns per unit-length of the coil. The direction of the magnetic field is along the axis of the coil and also follows the right-hand rule, with the thumb now being the direction of the magnetic field and the wrapping hand the sense in which the current is circulating around the coil. The pattern of the magnetic field lines outside the coil is similar to that of a magnet: The field lines leaving one end of the coil describe a loop that reconnects back into the other end of the coil. A fundamental but puzzling feature of nature is that the magnetic field lines are always closed loops because there are no magnetic monopole charges. A few elements, called ferromagnetic, are strongly magnetic. These can be common ones such as iron, nickel, and cobalt, or rare-earth elements such as neodymium and gadolinium. Using magnetic materials as cores of a coil greatly enhances the strength of the magnetic field of the coil. In this case, μ_0 is replaced by μ in Equation 1.13, where $\frac{\mu}{\mu_0}$ is of the order of a few thousand for most typical cores.

Coils can be made into useful devices, called electromagnets. The most fascinating applications are those used in junkyards for lifting iron-rich scrap metal. Electromagnets arise in electronics as well: Analog relays use an electromagnet in a clever design as a slave switch. Various types of relays exist, including solid-state ones, which serve the same function but operate under a different mechanism. Figure 1.6 shows a common type of analog relay. It has three contacts. The center contact, attached to a ferromagnetic material, is connected to only one contact when the relay is not energized. This is done by a spring mechanism. When the coil is energized, the magnetic field attracts the center contact overcoming the spring, disconnecting it from one contact and pulling it toward the other contact. Most high-power devices use relays. The "on" button does not directly close the power switch. Instead, the button energizes the relay that closes the actual switch. The electrical bolt-locks of doors at the entrance of apartment buildings also use electromagnets to move the bolt lock (the buzzing sound that you hear is the coil being energized).

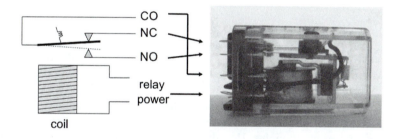

Figure 1.6. Diagram and photo of a relay. When power is not applied to the coil, the changeover contact (CO) is connected to the normally closed (NC) contact, and disconnected from the normally open (NO) contact. When power is applied to the coil, the CO contact is connected to the NO contact, and disconnected from the NC contact.

Many of us played with magnets when we were young: We approached a loose magnet with another we were holding to see the loose magnet flip and snap to the one in our hand. You can do the same thing with coils, one rotating and another fixed; if you are clever in switching the direction of the current in the rotating coil, you have a motor! The technology of making electric motors is mature, so you can see real works of engineering when you open one.

Playing with magnets also may have taught you to make magnets out of paper clips. With current-carrying wires, you can make ferromagnetic materials become permanent magnets and even switch the direction of their permanent magnetism. This is how magnetic tapes and computer disk drives store information (at least for now).

1.6.2 Inductors

Electromagnetism has more surprises: magnetic induction. If we put one coil against another one and *change* the current in one, it induces an electromotive force (a voltage) in the other coil *opposing* the change in flux. This phenomenon is called *magnetic induction*. The inductive effect is mediated by the change in magnetic field flux. You will later see that this induction from one coil to another is the mechanism behind a transformer, used to step up or down an alternating-current (AC) voltage or current. However, even a single coil carrying a current that changes experiences a back action called *self-induction*. This self-induction is quantified by the *inductance* of the coil. Since the flux depends on the area, geometry is an important factor in determining the inductance of a coil. For a cylindrical coil of length l and cross-sectional area A, and with no core, the inductance is

$$L = \mu_0 nAl \qquad (1.14)$$

Coils may have a core to increase their inductance. As circuit elements, self-inducting coils are known as *inductors*.

The effect of inductance on an individual inductor is an induced voltage as a function of the varying magnetic flux, which is generated by the current. Thus, we connect the induced voltage to the changing current: If there is a change in the current I flowing through the inductor, a voltage V appears across the terminals of the inductor opposing the change in the current. This voltage is given by

$$V = L\frac{dI}{dt} \qquad (1.15)$$

Induction is used in many applications. Later in this chapter you will see some specific circuits, but an interesting application is a special type of outlet labeled GFCI. These are often seen in bathrooms and kitchens (new building code regulations in the U.S. require them). When a coil senses a current imbalance in the wires of the outlet (a sign of trouble), it triggers the opening of a switch that turns off power to the outlet. Another interesting application is car-sensing stoplights. How do the stoplights sense your car has arrived at the intersection? It does not involve a camera. The car is sensed by an inductor placed under the road near the intersection. A car moving on top of the coil alters the inductance of the coil (acting like a core) and thus changes the current flowing through it. The change triggers the circuitry of the stoplight to switch the light to green. Inductors also play important roles in switching circuits that step up or down the voltage. The flash in a camera is an interesting example of all that we have talked about so far. The flash involves a short discharge of a large capacitor at a potential of 300 to 400 V. Switching circuitry and some trickery (precisely what we want to learn here) charges a capacitor to a high voltage from a few-volt battery pack. We discuss those circuits in Chapter 10 (Operational Amplifiers).

A new component, still under development, is being referred to as the "fourth fundamental electric component," together with the resistor, capacitor and inductor (quite a distinguished status!). This circuit element is the *memristor*. It is a resistor whose resistance depends on the rate of change of magnetic flux through it per unit charge flow: $R = \frac{d\Phi_B}{dq}$. This device is of interest because it has a resistance that depends on its value at a previous time—that is, it has memory. Since it is still under development we not discuss it further, but stay tuned, as you may hear more of it in the future.

1.7 POWER

The main processes behind electrical resistance are electron collisions within the material. In these processes, electrons collide and transfer part of their energy to the colliding partners. This energy goes on to heat the entire material. Consequently, a greater electrical current means a greater dissipation of energy. Indeed, hair dryers and toasters work on the heat generated by resistors that carry a large current. The warmth that we feel in all electronic boxes that have been on for a while (such as a radio or video recorder) is caused by resistive heating. As a consequence, heat dissipation is an important design consideration in building electronic boxes. Otherwise, the device overheats and fails (so do not place books over the vents of your video player).

The power dissipated by resistors is given by

$$P = IV = I^2 R = \frac{V^2}{R}$$ (1.16)

where we have used Ohm's law to provide different versions of the power dissipation formula. The unit of power is the watt (W). Care should be also taken when designing circuits so as to not exceed the power rating of the resistors. Normally we use $\frac{1}{8}$ and $\frac{1}{4}$ W resistors in the lab; they have the characteristic cylindrical shape of about 8 mm long and 2 mm in diameter. Table 1.4 gives the maximum ratings for different value $\frac{1}{4}$-W resistors. Most practical circuits use resistances of the order of 1 to 100 kΩ. Thus, using power supplies of up to 15 V is okay. Conversely, if we apply 15 V to a 100-Ω, $\frac{1}{4}$-W resistor, we would be dissipating 2.25 W. It would start smoking!

Table 1.4. Maximum Current and Voltage Ratings for 1/4-W Resistors

R (kΩ)	I_{max} (mA)	V_{max} (V)
0.5	22.4	11.2
1	15.8	15.8
10	5	50
100	1.6	158
1000	0.5	500

We must also take into account the power rating of the power supply. As a rule of thumb, you must multiply the maximum voltage and current readings on the scales of the supply. For example, the typical lab supply has 0 to 20 V and 0 to 0.5 A scales. Thus, the supply can handle 20·0.5 = 10 W. When you get close to using this much power you will feel a definite warmth in your supply.

High voltage supplies that go up to a few kilovolts have a maximum current rating of a few milliamps, so they are still somewhat safe. Be careful with supplies that put out more than 50 V and have a power rating greater than 50 W: They can cause electrocution. It is always important to know the dangers of the equipment that you use. Your body has a resistance of about 1 to 10 kΩ (arm-to-arm), and a current of at least 50 mA through the heart causes cardiac arrest. The wall outlet puts out an average of 110 V (in the U.S.; 220 V in many other countries), with a maximum average current of 10 to 30 A. That is deadly, indeed! Because of this, equipment cases are connected to earth ground. This is the third (or fourth) conductor in outlets, which is connected to a metal rod penetrating mother Earth (also a conductor) below the basement of the building. A live wire inside the box accidentally touching the metal chassis of the box will send the current to ground. Touching the box in this situation would not matter because charges prefer the easier path to ground. It is common to bypass electronic boxes (such as computers or players) with "cheater plugs" that cut the connection of the third grounding conductor of your plug. Sometimes this is necessary because certain cheap extension chords or outlets do not have the ground-conductor feature. However, they eliminate the safety feature of the electronics box: If you touch the case when there is a short, the charges will go through you to ground. The rule of thumb is that this is okay if the power to the box goes through a power adapter; they almost always reduce the input voltage to less than 15 V. *Never* use the cheater plug with high-power appliances (such as a washer or toaster) that are connected directly to the utility outlet, as you will be bypassing its electric-shock safety design, and risking your life.

Even the small dissipation in conductors can be significant when large currents are used, and they have to be cooled with water or other means. Otherwise, the conductors will melt! For example, the 24 wire-gauge that we use in the lab has a maximum current rating of 3.5 A. Because the melting of wires or circuit elements can produce a short with bad consequences (a fire), all properly designed circuits have a *fuse*. When a short circuit causes an over-current, the fuse, normally a metal that melts upon the passage of the over-current, breaks the circuit.

We mentioned a strange word earlier: noise. This is a technical term that refers to fluctuations in the electrical signals. In some cases, this is due to inherent physical properties of electrical components, such as resistors. Because electrons follow random walks as they go through a resistor, they introduce a fluctuation in the voltage drop across the resistor, called *Johnston noise*, which is proportional to $\sqrt{RT}$, with R and T being the resistance and the temperature, respectively. Noise produced by random fluctuations in electrical currents, called "shot noise," depends on $\sqrt{I}$. It is due to the quantized nature of the charge carriers and the fact that they go by at random times. These and other forms of noise become important when circuits rely on the stability of the electrical signals.

1.8   CIRCUITS

Enough of definitions and concepts. Let us get started with the main event of the course: circuits. We begin with elementary notions on how to connect resistors together, and follow quickly with circuits.

1.8.1   Equivalent Resistances

We can reduce the effect of a group of resistors connected together to the effect of a single resistor. We can classify the combination of resistors in a circuit as one of two types: in series and in parallel. If we have two resistors with values R_1 and R_2 connected back to back, or *in series* (Figure 1.7), the arrangement is equivalent to one resistor with a value equal to the sum of the values of the individual resistors:

$$R_{eq} = R_1 + R_2 \tag{1.17}$$

Figure 1.7. Resistances connected in series.

This can be generalized to more than two resistors. The equivalent resistance is the sum of the values of all the resistors in series:

$$R_{eq} = \sum R_i \tag{1.18}$$

For example, two resistors with the same value R connected in series are equivalent to one resistor with a value of $2R$.

When resistors are connected together to divide the current path into two or more paths, or connected *in parallel* (Figure 1.8), the equivalent resistance is found the following way: Its reciprocal is the sum of the reciprocals of the individual resistors. In the case of two resistors, it is

$$\frac{1}{R_{eq}} = \frac{1}{R_1} + \frac{1}{R_2} \tag{1.19}$$

For example, two resistors with the same resistance R connected in parallel have an

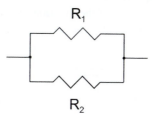

Figure 1.8. Resistors in parallel.

equivalent resistance of $\frac{R}{2}$. If the voltage is applied across the two resistors, then the current splits evenly through the two resistors. Another case involves calculating the equivalent resistance of a small resistor in parallel with a much larger resistor. The small resistor wins: The equivalent resistance is approximately the smaller resistance. Because both resistors have the same voltage across, by applying Ohm's law we get that the ratio of the currents through the resistors is equal to the inverse ratio of the currents:

$$\frac{I_1}{I_2} = \frac{R_2}{R_1} \tag{1.20}$$

An extreme case of this is a resistor and a straight wire (with "zero" resistance) in parallel: The equivalent resistance is "zero." We use quotes because a straight wire still has some very small resistance. Popular treatments describe this in a more qualitative way: The current takes the "easier" path. That is true, but it is more accurate to say that the current through the resistor is negligible compared to that of the wire. For example, a 1-inch piece of 24-gauge wire ($R = 0.13\ \Omega$) in parallel with a 1-kΩ resistor carries a current that is 7,500 times the current through the resistor.

We can generalize the formula for more than two resistors in parallel:

$$\frac{1}{R_{eq}} = \sum \frac{1}{R_i} \tag{1.21}$$

■ **EXERCISE 1.1**

Calculate the equivalent resistance of the network in Figure 1.9. All resistors have resistance R.

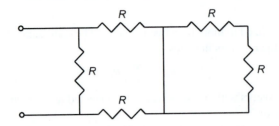

Figure 1.9. Circuit for Exercise 1.1.

1.8.2 Kirchhoff's Laws

Circuits are solved with two basic rules: the *junction rule* and the *loop rule*.

1. The junction rule states that the sum of all the currents flowing into a junction must be equal to the sum flowing out of the junction. We can label those currents going in as positive and those going out as negative. In the example of Figure 1.10, the junction rule becomes

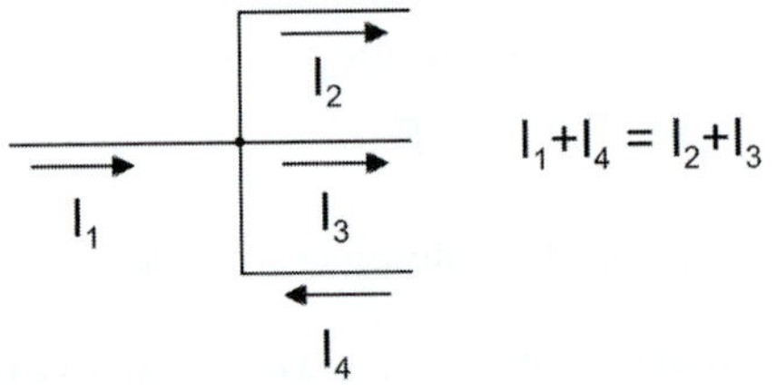

$$I_1 + I_4 = I_2 + I_3$$

Figure 1.10. Junction rule of circuits: The sum of currents into a junction is equal to the sum of currents out.

$$I_1 - I_2 - I_3 + I_4 = 0$$

The rule essentially reflects the conservation of charge. All charges go somewhere; they do not appear or disappear. In general, for any number of branches reaching a junction, the currents in all the branches must satisfy

$$\sum_{junction} I_i = 0 \qquad (1.22)$$

2. The loop rule states that the sum of all the voltage drops around a loop is zero. The circuit of Figure 1.11 has a voltage source and three resistors in series. First note that the source voltage V_0 (on the left) is represented by the symbol of a "single-cell" battery. This is an old-fashioned representation, but it suits our purpose. Because the circuit is closed, the current comes out of the source through the positive end (larger bar) and returns through the negative end (shorter bar). For the current to go through

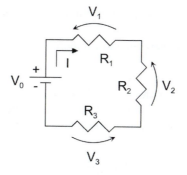

Figure 1.11. Loop rule of circuits: The sum of the voltages in a loop is zero.

each of the resistors, one side of each resistor must be positive and the other negative. We specify the potential drops across each resistor by V_1, V_2, and V_3. We can write an equation for the loop rule as:

$$V_0 - V_1 - V_2 - V_3 = 0$$

Replacing the voltage across each resistor by the product of the current and the resistance (given by Ohm's law) results in

$$V_0 - IR_1 - IR_2 - IR_3 = 0$$

Finally, the current flowing through the circuit can be found from the previous equation as:

$$I = \frac{V_0}{R_1 + R_2 + R_3}$$

Notice that the denominator does not distinguish whether we have three resistors or one with an equivalent resistance of $R_1 + R_2 + R_3$ (resistors in series). In summary, the voltages across elements in a loop must satisfy

$$\sum_{\text{loop}} V_i = 0 \tag{1.23}$$

1.8.3 Voltage Dividers

Most electrical circuits process information in the form of voltages, as illustrated in Figure 1.12. The utility of a particular circuit stems from the relationship between the

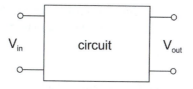

Figure 1.12. The essence of most electronics circuits: to transform an input voltage into an output voltage.

input and the output voltages. Another way to put it is to say that the output is a processed version of the input.

A voltage divider is the simplest and most important type of voltage processor. It uses two resistors to generate a voltage that is the input divided by some factor. Its importance rests in the fact that most electronic circuits can be understood in terms of voltage dividers. If you understand the voltage divider, you understand *a lot* about electronics. Figure 1.13 shows the schematic of a voltage divider. Let us assume that V_{in} is generated by some voltage source. To get V_{out} from V_{in}, we first find the current following Ohm's law:

$$I = \frac{V_{in}}{R_1 + R_2} \tag{1.24}$$

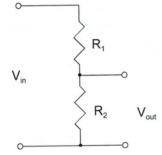

Figure 1.13.   A voltage divider.

Then applying Ohm's law again across R_2, we get

$$V_{out} = IR_2 \tag{1.25}$$

Replacing I from Equation 1.24 into Equation 1.25 we get

$$V_{out} = \frac{R_2}{R_1 + R_2} V_{in} \tag{1.26}$$

A consequence of this is that $V_{out} < V_{in}$. You may be concerned that we are ignoring what V_{out} is connected to. We will consider the matter in detail in Chapter 6 (AC Signals). The main point of the voltage divider is that we use resistors to divide an input voltage. This not the whole story, though. If we connect a resistor to V_{out}, then for V_{out} not to change significantly, the resistor should be *much larger* than R_2. The voltage divider concept is not so much a way to generate smaller voltages from a larger one, but a way to quickly figure out the voltages across circuit components, as opposed to using the loop rule.

■ **EXERCISE 1.2**

Use the concepts of resistors in series and parallel with a voltage divider to calculate the voltage across each resistor in the circuit of Figure 1.14. Each resistor has a value of 1 kΩ.

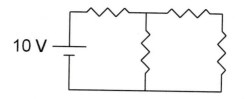

Figure 1.14. Circuit for Exercise 1.14.

1.8.4 Multiloop Circuits

To determine the currents flowing in all branches of a circuit, we have to solve as many equations as the circuit has branches. The standard procedure is:

1. Draw the circuit diagram and label currents for each branch of the circuit. A branch is any piece of circuit between two junctions. Assume also a likely direction for the currents. Do not worry if you are not sure. If any answer comes out negative, it means that the current flows in a direction opposite to the one you assumed.

2. Apply the loop rule to all loops. The direction of the currents that you chose specifies the direction of the voltage drops.

3. If there are k joints in a circuit, there are only $(k-1)$ independent junction equations. Therefore, apply the junction rule to $(k-1)$ joints.

4. Solve the system of junction and loop equations.

For example, suppose we want to find the currents flowing through each resistor in the circuit of Figure 1.15. Assume that the values of the three resistors, and V_1 and V_2 are known. The circuit has three branches and, consequently, two loops. We assume the direction of the currents I_1, I_2, and I_3, and from the two loops, we get

$$V_1 - I_1 R_1 - I_3 R_2 = 0$$

$$I_3 R_2 + I_2 R_3 - V_2 = 0$$

From one of the two joints, we get

$$I_3 - I_1 - I_2 = 0$$

The previous equations constitute a system of three algebraic equations with three unknowns, which is straightforward to solve.

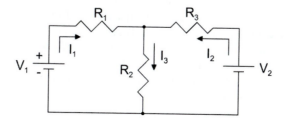

Figure 1.15. Solving a multiloop circuit (see text).

Let us do a numeric example using Figure 1.16(a). Our task is to find the currents going through each resistor. We make assumptions on the direction of currents, as shown in Figure 1.16(b), which then imply the potential drops shown.

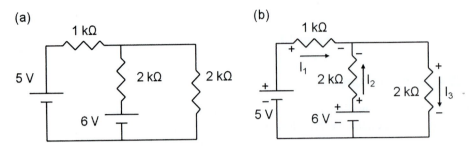

Figure 1.16.   An example of a multiloop circuit (see text).

Applying the loop rule to the loop on the left, we write

$$5V - I_1 \cdot 1k\Omega + I_2 \cdot 2k\Omega - 6V = 0$$

Applying the loop rule to the loop on the right, we write

$$6V - I_2 \cdot 2k\Omega - I_3 \cdot 2k\Omega = 0$$

The outer segments form an outer loop, but we do not need it. It would yield an equation that is redundant with the other two. We have two joints, so we use only one:

$$I_1 + I_2 - I_3 = 0$$

We can solve these equations in a straightforward way. For example, in the last equation, you put I_3 in terms of I_1 and I_2 and substitute it into the second equation. You then have two equations with I_1 and I_2 that you can solve, and get $I_1 = 1\,mA$ and $I_2 = 1\,mA$. Replacing the value of these currents in the last equation, you get $I_3 = 2\,mA$. In the last example, we could have avoided the units, using V for voltage, $k\Omega$ for resistance, and mA for current, because $mA = V/k\Omega$.

1.8.5 Transient Circuits

In previous circuits, the currents and voltages remain constant once the circuit is connected. In this section we consider how currents and voltages reach their steady-state values immediately after the circuit is connected, and how those steady-state values reach zero when the circuit is disconnected. We do so using capacitors and inductors because all circuit elements (resistors and even wires) have small capacitive and inductive properties. The deliberate use of capacitors and inductors in our discussion allows us to understand the transient values that the currents and voltages take when the steady-state of the circuit is changed. We treat capacitive and inductive effects separately, but we should keep in mind that a real circuit will have both.

Capacitive   Circuits with capacitors are somewhat tricky because capacitors are basically open circuits. The action occurs when the capacitor is charging or discharging. In a steady state situation where the current through the circuit is constant, capacitors do not contribute much; they remain charged by a constant amount, according to the potential across them.

Since the current is the flow of charge, the current going in or out of a capacitor must be the rate of change of the charge in the capacitor. Thus, to obtain it, we must differentiate the capacitor equation $q = CV$:

$$I = \frac{dq}{dt} = C\frac{dV}{dt} \tag{1.27}$$

Consider the circuit shown in Figure 1.17. When the switch is in position 1, the capacitor gets charged so that it has a voltage V_0. When we flip the switch to position 2, the capacitor discharges through the resistor. Because the two are in parallel, the voltage across the capacitor is $V = IR$. Replacing I into Equation 1.27, we get

$$V = -RC\frac{dV}{dt} \tag{1.28}$$

The negative sign accounts for the fact that the voltage is decreasing as the capacitor gets discharged. The solution to Equation 1.28 is obtained by rearranging both sides and integrating the time from 0 to t and the voltage from V_0 to V. The result is

$$V = V_0 e^{-t/RC} \tag{1.29}$$

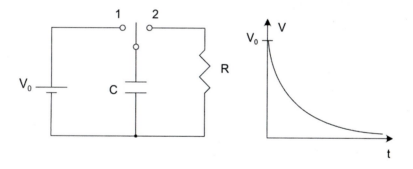

Figure 1.17. Circuit to study the discharge of a capacitor (left), and the graph of the resulting voltage across the capacitor as a function of time (right).

The voltage across the capacitor decreases in a characteristically exponential manner, taking a time $t = RC$ to decay to $\frac{V_0}{e}$. Because exponents must not have units, notice the new relation between electrical units: ohms × farads = seconds.

■ **EXERCISE 1.3**

A capacitor is charged to 200 V.

1. *What capacitance does it have if it stores a charge of 10 mC?*

2. *If we connect a 1 kΩ resistor across the capacitor, what is the* maximum *current that will flow as the capacitor discharges?*

We can use a similar circuit to charge a capacitor *through* a resistor. This is shown in the circuit of Figure 1.18. When the switch is in position 2, the ends of the capacitor are connected directly to each other, and the capacitor is discharged. When the switch is set to position 1, the capacitor charges through the resistor. The voltage across the capacitor is

$$V = V_0 - IR \tag{1.30}$$

This equation encompasses the dynamics of the circuit in an interesting way. At an initial time when the capacitor is discharged, $q = 0$, so the voltage across the capacitor is zero. Current flows into the capacitor to charge it, creating a voltage drop across the resistor IR. When the capacitor is initially discharged, $V = 0$, so then from Equation 1.30, the current must be maximum: $I = \frac{V_0}{R}$. At the other extreme (after a long time), when the capacitor is fully charged, $V = V_0$, and so $I = 0$. The resistor and capacitor complement each other in an very intriguing way. If we substitute the definition of the current charging the capacitor, Equation 1.27, into Equation 1.30 we get

$$V = V_0 - RC\frac{dV}{dt} \tag{1.31}$$

The solution of this equation is

$$V = V_0(1 - e^{-t/RC}) \tag{1.32}$$

You can see from the solution that $V = 0$ at $t = 0$, and when $t \rightarrow \infty$, the voltage approaches V_0. The graph of Figure 1.18 shows this.

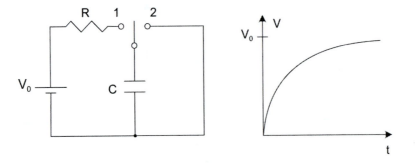

Figure 1.18.   Circuit to study the charging of a capacitor (left), and a graph of the resulting voltage across the capacitor as a function of time (right).

We can check our initial comment about the current by calculating the current I:

$$I = \frac{dq}{dt} = C\frac{dV}{dt} = \frac{V_0}{R}e^{-t/RC} \tag{1.33}$$

The previous equation implies that the current is maximum at $t = 0$, with $I_{max} = \frac{V_0}{R}$. We can also see that it is zero when $t \rightarrow \infty$. That is, when the capacitor is fully charged, it acts as an open circuit and no current flows in the circuit. In the first circuit, we charged the capacitor by connecting it to the power supply V_0. There the charging time was also finite but very short because the "resistor" was the resistance of the wires (of the order of milliohms).

Earlier we talked about the effect of the capacitance of dangling wires: If we change the capacitance, this creates a current to redistribute the charge so that $V = \frac{q}{C}$. This charging/discharging current produced by vibrations jiggling the wires causes microphonics.

Notice also that it takes several *RC* time constants to charge or discharge a capacitor. Electronic switching circuits have to be designed with care so that *RC* time constants do not limit their switching speed. In digital circuits, discussed in the first part of the course, capacitors play a supportive role: They stabilize the voltage supply. Consider the following situation. Suppose that we connect a circuit to a voltage supply that is generated from the electrical outlet. Suddenly, a spike in the voltage occurs. This can be caused by a number of reasons: thunderstorms or when a high-current circuit connected to the same outlet is switched on and off. The reason is not important. These spikes are common in circuits, and all devices must protect themselves against them. Capacitors are components that help with this. If we are not protected, the voltage spike could alter the digital value of the signals in a circuit, introducing a bogus error and malfunction. If we put a capacitor in parallel with the voltage supply, the surge in charge caused by the power spike would go on to charge the capacitor instead of going through the circuit. If you look at a computer board, you will see brightly colored round components next to each integrated circuit; they are capacitors protecting the device from spurious surges.

Inductive   Inductors play an important role in transient circuits. Let us analyze the circuit of Figure 1.19. Upon connecting the switch to position 1 the battery is connected to a circuit where a resistor *R* is in series with an inductor, which we label by *L*. Writing down the loop rule for this circuit gives us

$$V_0 - IR - L\frac{dI}{dt} = 0 \tag{1.34}$$

This is an integral equation similar to the one for the capacitor, except that here we solve for the current. The initial conditions are: $I = 0$ at $t = 0$. By applying our calculus tricks and integrating for I at time t, we get

$$I = \frac{V_0}{R}(1 - e^{-Rt/L}) \tag{1.35}$$

The graph of this result is similar to the one for the voltage in Figure 1.18. Notice that after a long time, the current is $\frac{V_0}{R}$; or what we would get if we connect the supply across the resistor. That is, the inductor slows the rise of the current. The rise depends on a new time constant: $\frac{L}{R}$. If we switched the voltage of the supply suddenly to zero (changing the switch to position 2), the current would decrease exponentially with the characteristic time constant, giving a time dependence similar to voltage across the capacitor in Figure 1.17. In high-frequency applications, even individual wires act like inductors, so short wires and other precautions are needed to decrease the time constants. Otherwise, the inductive aspect of the circuit will not let the circuit switch back and forth from high to low voltages, a vital requirement of digital circuits. The same types of circuits are used in surge protectors to

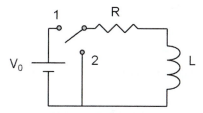

Figure 1.19.   Circuit to study the current through a circuit with an inductor.

protect sensitive devices from current spikes in the line. Inductors complement the job that capacitors do in stabilizing power supplies.

Inductors are a vital component of fluorescent lamps. The popularity of these sources stems from their high efficiency. Figure 1.20 shows a schematic of the wiring diagram of a simple fluorescent tube. The lamp works by creating an electric discharge in a low-pressure tube. The gas is typically argon with mercury in it. The ultraviolet mercury emissions excite the phosphor inner coating of the tube, which fluoresces in the visible portion of the spectrum of light. Once a discharge has been established, the line voltage can maintain the discharge. However, the voltage required to get the discharge started is much higher. To provide an initial high voltage, an inductor ballast is used in series with a starter switch. When the switch opens, the inductor generates a high voltage that ignites the discharge. The lamps have additional filaments at either end to preheat the gas and ease the initial discharge.

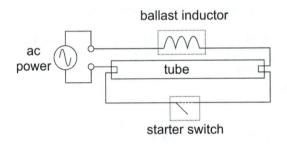

Figure 1.20. Circuit for a fluorescent tube.

Let us analyze this application in more detail with the circuit of Figure 1.21. A closed switch allows the passage of current $I = \frac{V_0}{R}$ through the circuit. At this point the inductor plays no role. To initiate the process, we open the switch that is in parallel with the discharge lamp, of resistance R_l. The resistance of the circuit suddenly changes, so the current changes, kicking the inductor into action. The loop rule for the circuit is now

$$V_0 - IR - IR_l - L\frac{dI}{dt} = 0 \qquad (1.36)$$

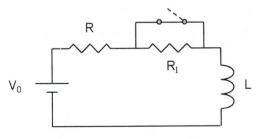

Figure 1.21. A circuit to analyze the startup of the fluorescent tube.

Notice the sign of the last term: The inductor wants to keep the current flowing, so the negative sign together with the $\frac{dL}{dt} < 0$ gives a net *positive* contribution. The solution to this equation is slightly more elaborate than the one of the previous circuit, but it yields

$$I = \frac{V_0}{R+R_l}\left[1-\left(1+\frac{R+R_l}{R}\right)e^{-(R+R_l)t/L}\right] \tag{1.37}$$

We can verify its consistency by noticing that at $t=0$, we have that $I=\frac{V_0}{R}$. What is the big deal? Notice that the voltage across the load is $V_0\frac{R_l}{R}$; it is the supply voltage multiplied by the factor $\frac{R_l}{R}$, which by design is much larger than 1. What is responsible for this? The culprit is the inductor. If you calculate the voltage across the inductor, you get

$$V_L = \frac{V_0 R_l}{R}e^{-R_l t/L} \tag{1.38}$$

where we have done $R_l + R \sim R_l$ to simplify the algebra. A key component is that a large L prolongs the transient voltage. After several time constants, the current decays down to $\frac{V_0}{R_l}$ (remember $R_l \gg R$).

   This same circuit is used in all sorts of practical joke devices. For example, an innocent victim is set up to open a book with a suggestive title (well, the victim is not so innocent), which is a trap. The book cover is conspicuously metallic. As the victim holds the book with one hand and opens the cover with the other hand (a switch in disguise), you hear "whoosh" and "clang" (the victim gets the surprise jolt, swings the book goes up in the air, and vehicle of evil lands not so softly on the ground). You might guess what is R_l: the victim. These tricks do not last long, as they can only withstand so many hard crashes! There are many other practical jokes like this one, all using the same principle: induction.

1.9   ABSTRACTIONS AND SYMBOL JARGON

Before closing our introduction, we must discuss a few simplifications that are done in electronics for the sake of clarity. A circuit always involves a closed loop. When a complex circuit has several subcircuits that draw from the same power supply, it becomes cumbersome to draw all the wires to the closed loop. On one end, all the circuits are connected to the positive side of the supply, and on the other end, to the negative side of the supply. Figure 1.22 shows a simplification of the connections to the supply. An arrow pointing up represents the connection to the positive end. If the voltage is negative the arrow usually points down. The other side of the supply is the "common," and is represented by an inverted triangle. In many circuits this common line is also chassis ground, or the protective connection to earth-ground.

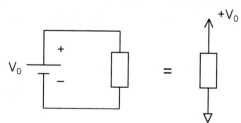

Figure 1.22.   The jargon of electronics circuits: The diagram circuit on the left is simplified by the one on the right.

   How do we draw a circuit diagram? In this regard, we are going to be "square," which means that all wires are either horizontal or vertical. Only in exceptional cases do we use diagonals, and *never* use squiggly lines. A circuit diagram is abstract and dense, so we need clarity to understand it. A clear and neat diagram is easy to understand. A cluttered one with wires going everywhere they please is a headache.

We also have a system to label if wires are connected. If a wire terminates on another wire, as shown on the left side of Figure 1.23, we assume that the two wires are connected—what would be the point of terminating the wire and leaving it unconnected? Sometimes, to be reassuring, we put a black dot at the intersection "T," but this is overkill. We use both notations interchangeably in this book.

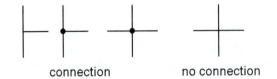

connection   no connection

Figure 1.23.   Convention for use in drawing circuit diagrams.

When two lines cross, we could have a dilemma: Are they connected or not? To distinguish the two cases we put a black dot on the cross to denote that the lines are connected, as shown in Figure 1.23. This is important because the lack of a dot means that the lines have nothing to do with each other, and we train ourselves to ignore those crossings when following lines in a circuit diagram.

1.10 PROBLEMS

1. Two capacitors with capacitances $3\,\mu F$ and $2\,\mu F$ are initially discharged. They are connected in series, and then the two ends of the combination are connected to a 10-V battery, as shown in Figure 1.24. The negative end of the battery is at zero potential.

   (a) How much charge does the battery deliver? You can get started by figuring out the relationship between the charges in each capacitor and the voltage across each of them.

   (b) If we connect a single capacitor to the same battery, what would be the capacitance of the capacitor so that it draws the same charge from the battery?

   (c) Find the charges on each capacitor.

   (d) Find the voltage across each capacitor.

   (e) If point B is at zero potential, what is the potential at point A between the two capacitors?

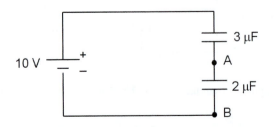

Figure 1.24.   Circuit for Problem 1.

2. A battery with potential V_0 is connected to a capacitor C. We label the charge in the top plate as q_1 and the charge in the bottom plate as q_2. Which statement is correct?

   (a) $q_1 = q_2 = CV_0$
   (b) $q_1 = -q_2 = CV_0$
   (c) $q_1 + q_2 = CV_0$
   (d) $q_1 = -q_2 = 2CV_0$
   (e) $q_1 = q_2 = C\frac{V_0}{2}$

3. Suppose that we have two capacitors with capacitances $0.1\ \mu F$ and $0.2\ \mu F$. We connect them in parallel and apply a voltage of $10\ V$ to their ends.

   (a) What is the charge on each capacitor?
   (b) What is the total amount of charge that the source has to supply to charge the two capacitors?
   (c) If we replaced both capacitors with one so that it draws the same amount of charge from the supply, what capacitance would it have?
   (d) Can you generalize the previous situation for finding the equivalent capacitance of any two capacitors connected in parallel?

4. In the circuit of Figure 1.25, $C_1 = 0.2\ \mu F$ and $C_2 = 0.15\ \mu F$. When a voltage of $10\ V$ is applied to the system of capacitors, we find that the charge of capacitor 1 is $1\ \mu C$.

   (a) Find the value of C_3, and the charge on it.
   (b) What is the equivalent capacitance of the arrangement?

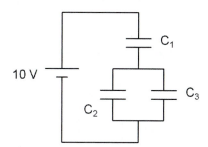

Figure 1.25.   Figure for Problem 4.

5. We apply $1.2\ V$ to the $0.5\ \mu F$ capacitor of Figure 1.26. We then disconnect the capacitor from the power supply. Subsequently, we connect an uncharged $1\text{-}\mu F$

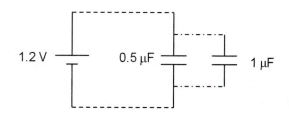

Figure 1.26.   Figure for Problem 5.

capacitor in parallel with the other capacitor. Find the charge on the 1-μF capacitor. (Note that the voltage across it is not 1.2 V.)

6. The potential of point A in Figure 1.27 is 0 V. Find the potential of point B.

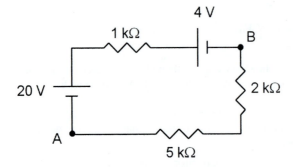

Figure 1.27.   Figure for Problem 6.

7. The colors of the bands of the resistors of Figure 1.4 are: (a) orange, orange, red, white; (b) brown, black, red, gold; (c) brown, red, blue, orange, green.

   (a) Determine their value and tolerance.
   (b) A measurement of the actual resistance of (b) gave 992 Ω. Should we return it as defective? Explain.

8. Calculate the equivalent resistance of the network in Figure 1.28.

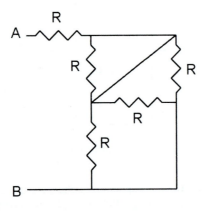

Figure 1.28.   Circuit for Problem 8.

9. If point A is at zero potential in Figure 1.29, what is the potential at points B and C?

10. Many of the circuits that you will use in the lab will be powered by +12 V power supplies. What is the smallest value $\frac{1}{8}$-W resistor that we can apply the full voltage of the power supply without burning it?

11. Be aware of the dangers of electricity. Human skin can exhibit large variations in electrical resistance. Although dry skin may have a resistance of 100 kΩ, wet and tender skin may have resistances as low as 1 kΩ. If electrocution is caused by currents

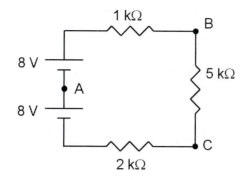

Figure 1.29.   Circuit for Problem 9.

above 50 mA, what applied voltages would cause electrocution for (a) dry and (b) wet skin?

12. Using *only* the concepts of equivalent resistance and voltage divider, calculate the voltage between points A and B of Figure 1.30. Hint: First find the voltage drop across the 8 k resistor, but do not ignore the resistor ladder to the right of it.

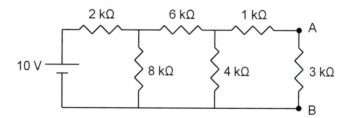

Figure 1.30.   Circuit for Problem 12.

13. Find the value of R in the circuit of Figure 1.31 if points A and B are at the same potential.

14. Two resistors are connected in series to a 20 V battery. The ratio of their resistances is 1:4.

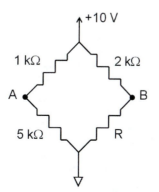

Figure 1.31.   Circuit for Problem 13.

(a) Find the voltage across the larger resistor.

(b) If the total current flowing through the resistors is 10 mA, what is the value of the smaller resistor?

15. Find the current flowing through each resistor, and the voltage drop across the resistor in the middle branch of the circuit in Figure 1.32.

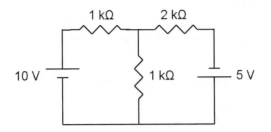

Figure 1.32.   Circuit for Problem 15.

16. Find the current flowing through the 4-kΩ resistor in Figure 1.33.

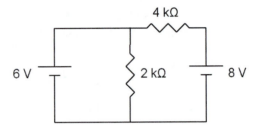

Figure 1.33.   Circuit for Problem 16.

17. Find the value of the resistors R_1 and R_2 in Figure 1.34.

18. The capacitor in the circuit of Figure 1.35 is discharged.

(a) Switch 1 is closed at $t = 0$. At what time will the voltage across the capacitor reach 6 V?

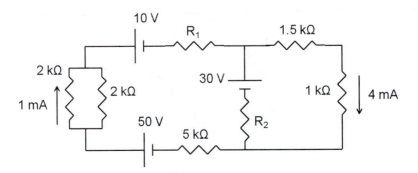

Figure 1.34.   Circuit for Problem 17.

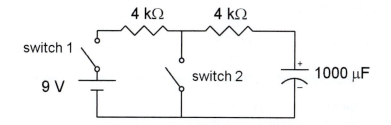

Figure 1.35.   Circuit for Problem 18.

(b) Switch 1 is opened when the voltage across the capacitor reaches 6 V. What was the current flowing through the circuit before the switch was opened?

(c) We close switch 2. At what time after closing the switch is the voltage across the capacitor equal to 3 V?

19. At $t = 0$, the switch of Figure 1.36 is closed. At $t = 80$ ms, the voltage across the capacitor is 5 V. Calculate V_0.

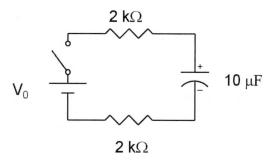

Figure 1.36.   Circuit for Problem 19.

20. In Figure 1.37, the capacitors are initially charged. As soon as the switch is closed the current is 0.5 mA. After 14 ms the current has dropped to 0.25 mA. Find C.

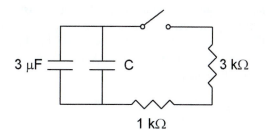

Figure 1.37.   Circuits for Problem 20.

1.11 LAB PROJECTS

For explicit lab instructions, see the *Lab Manual*.

1.11.1 An Application of the Voltage Divider: A Darkness Sensor

Required equipment: Power supply, multimeter.
Required components: Cadmium photoresistor, CD4066 digital switch, LED, resistors.

1. Suppose that we want to automatically turn on the lights of some place as soon as it gets dark. The circuit must consist of a light sensor and an electronic switch. The sensor tells the switch to close when it is dark, and the switch connects power to the lights (or whatever else) you want on. Figure 1.38 shows the circuit.

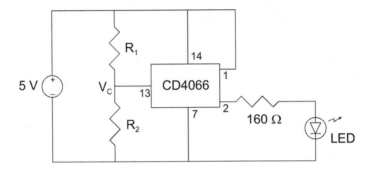

Figure 1.38.   Circuit for Lab project 1.11.1.

2. **Photoresistor**. The light-sensitive device we will use is a photoresistor. It has a resistance that varies with the intensity of the light that it receives. Place the photoresistor in the breadboard and measure its resistance for both light and dark situations.

3. **Electronic switch**. Our electronic switch is an old-technology CMOS analog IC switch (note that the 74HC4066 will not work). It works the following way: when $V_C > 2.5$ V a voltage applied to V_{in} gets connected to V_{out}. The IC needs to be powered by $+5$ V. The pins of the IC are numbered in a counterclockwise order, with pin 1 being on the top left (the top is defined by a small circular mark in the IC). To have something to turn on we have included a light emitting diode (LED). Therefore, if we apply more than 2.5 V to V_C the circuit for the LED gets switched on and it will emit light.

4. **Voltage divider**. You must design a voltage divider consisting of R_1 and R_2, with one of them being the photoresistor and the other one a resistor whose value you have to calculate. The operation of the circuit is such that when there is light, the voltage across R_2 is less than 2.5 V, and when it is dark the voltage is greater than 2.5 V. Try it before connecting it to the IC.

1.11.2 Delayed Switch

Required equipment: Power supply.
Required components: SPDT switch, CD4066 digital switch, LED, resistors, capacitors.

Often we want to take a group photo, but we want everybody in the picture. A delayed shutter is the solution. Here we use a circuit similar to the previous one, but we need to delay the closing of the switch by 2 seconds (a fast run to join the group). Check the schematic of the circuit in Figure 1.39. You must design a capacitor-charging circuit that triggers the switch ($V_C > 2.5\ V$) 2 seconds after the manual switch is closed.

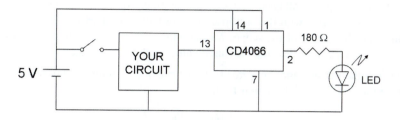

Figure 1.39.   Circuit for Lab project 1.11.2.

1.11.3   RC Circuit as an Integrator and Differentiator

Required equipment: Function generator, oscilloscope.
Required components: Resistor, capacitor.

1. This circuit requires that you use the oscilloscope and the function generator. You should become familiar with both. Try observing square and triangular waveforms, and vary their amplitude and frequency.

2. Assemble the circuit of Figure 1.40 using the function generator as the source. The circuit has a resistor in series with a capacitor. Using the loop rule, we get

$$V(t) - V_R - V_C = 0 \tag{1.39}$$

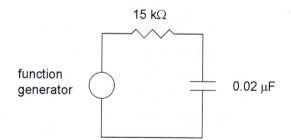

Figure 1.40.   Circuit for Lab project 1.11.3.

Because $V_R = IR$, $I = C\frac{dq}{dt}$ and $q = CV_C$, then the following equation is true:

$$RC\frac{dV_C}{dt} = V(t) - V_C \tag{1.40}$$

The condition $V_C \ll V$ must be satisfied to make the capacitor behave as an integrator, so we drop V_C in Equation 1.40. This is achieved whenever $T \ll RC$, where T is the period of the input voltage. Solving for V_C, we get

$$V_C = \frac{1}{RC} \int_0^{T/2} V(t)dt \tag{1.41}$$

Because $V(t)$ is constant during half a period, you can obtain a relationship between both peak-to-peak voltages. Verify the integrating properties of the above circuit for square waveforms with the oscilloscope. For better comparison, display V_C and V in channels 1 and 2 of the scope. Measure the peak-to-peak voltage and period (T) of both waves. Compare the measured values with the calculated ones.

3. Switch the resistor and capacitor from the previous circuit. Suppose now that we take the derivative of Equation 1.39:

$$\frac{dV(t)}{dt} - \frac{dV_R}{dt} = \frac{dV_C}{dt} \tag{1.42}$$

If $\frac{dV}{dt} \gg \frac{dV_R}{dt}$, and if we make $V_C = \frac{q}{C}$, then $\frac{dV_C}{dt} = \frac{I}{C}$. The voltage across the resistor is

$$V_R = RI = RC\frac{dV(t)}{dt} \tag{1.43}$$

That is, the voltage across the resistor is proportional to the *derivative* of the input voltage. The condition for this is met when $\omega_{source} \ll \frac{1}{RC}$.

Connect the function generator (set to triangular waveform) at a frequency that satisfies the previous condition, and observe and measure the resulting waves with the scope. Compare your measurements of V_R with the value you expect from the calculation. Without changing anything else, switch the waveform setting of the function generator to a square wave. Record your observations. Is this consistent with what you expected?

1.11.4 Practicum Test

Wire two 5 V power supplies in series. Connect them to two resistors in series with the following color bands: $R_1 =$ green, black, red, and $R_2 =$ brown, black, orange.

1. Measure the voltage across R_2 and compare it to your prediction.

2. Measure the current flowing through the circuit and compare it to your prediction.

3. Explain the differences between the predictions and the measurements.

4. Wire a capacitor and a resistor in series with the 5 V supply. Pick the resistor and capacitor so that $RC = 1$ s. Can you verify that your choice of components was correct?

PART I

DIGITAL

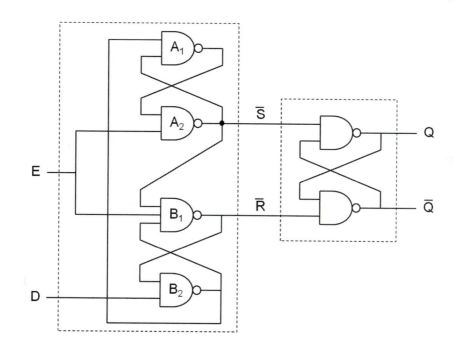

CHAPTER 2

INTRODUCTION TO DIGITAL ELECTRONICS

Contents

When we think about information in electrical signals the first thought that comes to mind is the magnitude of a voltage or a current. In an audio signal for example, higher amplitude means higher volume; that is, the actual magnitude of this signal conveys the information. This is true only in the realm of analog electronics. Digital signals are also represented by voltages or currents, but they convey information in a different way. The magnitude of digital signals can have only two values, which, for simplicity, we label as 0 and 1. The information in digital signals lies in the *combination* of these binary values, a binary number. In the audio example, a higher volume is specified by a higher binary number encoded in several electrical signals set to values 0 or 1.

Electrical noise plagues electrical signals. For example, the hissing that you hear when the stereo is cranked up all the way with no music playing is the result of electronic noise. This noise, due to thermal effects (among others), makes the value of an electronic signal jiggle to higher and lower values about some average value. The hissing sound is the jiggling of the electronic signal about zero volume. Analog recordings, such as those coming from tape recorders (now virtually obsolete), have a prominent background hiss that is present regardless of the volume. Analog radio (which still exists) suffers from the same problems. Digital signals are different. Because the values of the electronic magnitudes representing 0 and 1 can be very distinct, digital signals are much more immune to noise than analog signals. That is, it takes a lot to confuse, say, 0 V (a logic 0) with 5 V (a logic 1). Thus, the output of a digital recording (a CD or digital radio) is much cleaner, and suffers from hissing to a much lower degree than an analog recording. Some hiss is always present because the final stages of processing—the amplifier and speakers— are inevitably analog. In addition, the combination of values provided by digital signals, a code, can readily be used for computation (or cheating—MP3 is not a perfect reproduction).

Digital is most efficient for transmitting electronic information. The case of voice transmission in telecommunications is an excellent example of the distinction between digital and analog signals. The voice pattern that a microphone produces can be represented by a voltage as a function of time. As this signal is transmitted through a cable, it gets degraded by noise and distortion (a change in the shape of the signal due to non-ideal electronic components). By the time the signal gets to the other end of the transmission line, it no longer accurately represents the original signal (lots of hiss and a voice that seems to come out of a box). On the other hand, if before transmission we convert the information to binary, encoding the value of the incident voltage at each instant into a binary signal, and transmit those values in the form of a train of binary signals, we find that noise and distortion in the binary levels rarely result in a change of the encoded information. That is, signal degradation must be very extreme to confuse the binary electronic levels. When the signals reach the receiving end, the digital codes are converted back to analog voice signals that a speaker outputs.

Since digital electronics deals with only two values, the challenge in working with it resides in the encoding and processing of binary values. Because of this, we do not need to worry too much about the way in which the signals are represented and transmitted (such as voltage current levels). Thus, we postpone the study of analog electronics until the second part of the course. We start the course with a discussion of binary signals: how we make them and how we process them. Eventually, we will need to leave our digital cocoon and deal with the real analog world. That comes during the second part of the course. If you get into the habit of ignoring voltages and currents in the digital electronics part, we will take care of fixing that in the second part.

We start our discussion of digital electronics with a summary of concepts: binary numbers and digital functions. Toward the end of this chapter, we get a bit practical: What electronic devices do we use to process digital signals? This is still electronics, so we cannot ignore the analog part completely. Thus, at the end of the chapter we go over the voltages that we use to encode digital signals and other practical information. In subsequent chapters, we cover the various aspects of processing the digital signals.

2.1 NUMBER SYSTEMS

A number in base 10, such as 513.28, can be represented formally as

$$513.28 = 5 \times 10^2 + 1 \times 10^1 + 3 \times 10^0 + 2 \times 10^{-1} + 8 \times 10^{-2} \qquad (2.1)$$

In general, a number $N_r = N_3 N_2 N_1 N_0 . N_{-1} N_{-2}$ in base r is represented as

$$N_r = N_3 \times r^3 + N_2 \times r^2 + N_1 \times r^1 + N_0 \times r^0 + N_{-1} \times r^{-1} + N_{-2} \times r^{-2} \qquad (2.2)$$

The digits N_i can have values that range from 0 to $(r-1)$. Table 2.1 shows the first 20 numbers in four popular base systems.

Table 2.1. Number Systems

Decimal	Octal	Binary	Hex
0	0	00000	0
1	1	00001	1
2	2	00010	2
3	3	00011	3
4	4	00100	4
5	5	00101	5
6	6	00110	6
7	7	00111	7
8	10	01000	8
9	11	01001	9
10	12	01010	A
11	13	01011	B
12	14	01100	C
13	15	01101	D
14	16	01110	E
15	17	01111	F
16	20	10000	10
17	21	10001	11
18	22	10010	12
19	23	10011	13
20	24	10100	14

In the left column is our familiar decimal (base 10) number system. In the second column, we have the same number, but in the octal (base 8) system. This system was popular in the old days of 12-bit computers, but it is now obsolete. The third and fourth columns list the numbers in base systems that are useful in digital: binary (base 2) and hexadecimal (base 16). Binary is the number system of computers. As you will see in the next sections, hexadecimal digits are useful because they represent binary numbers in a compact way. The fractions of one hour (in minutes and seconds) that we use in time keeping are expressed in the sexagesimal system (base 60). This is a legacy of the Babylonians, from almost 4,000 years ago (1900–1650 BCE)!

NOTATION ALERT: When representing decimal numbers, we do not label the base. Otherwise, the base will be explicitly labeled as a subscript.

You can see in the table that in each number system of base r there are r different characters, including zero. Because 16 characters are needed for hexadecimal, we use the first five characters of the Roman alphabet for the digits after 9: $A_{16} = 10, B_{16} = 11, C_{16} = 12, D_{16} = 13, E_{16} = 14,$ and $F_{16} = 15$. A couple definitions are important to mention at this point: One binary digit is called a *bit*, and a set of eight binary digits is called a *byte*.

2.1.1 Number-System Conversions

The existence of different number systems forces us to worry about converting from one to the other. In the following sections, we discuss several "easy" conversions between number systems.

Any Number System to Decimal   To convert any number system to decimal, apply the definition of Equation 2.2. For example:

$$1101.11_2 = 1 \times 2^3 + 1 \times 2^2 + 1 \times 2^0 + 1 \times 2^{-1} + 1 \times 2^{-2} = 13.75$$

or

$$2A_{16} = 2 \times 16^1 + A \times 16^0 = 32_{10} + 10_{10} = 42$$

■ **EXERCISE 2.1**

What are the decimal representations of 33_8 and 33_{16}?

You can verify that the entries in Table 2.1 agree with this definition. The conversion back is more elaborate and is discussed in the next section.

■ **EXERCISE 2.2**

What is the value of $1F_{16}$ in decimal?

Four-Digit Binary to Decimal   This conversion is easy to do. We need to remember, that the "least significant bit" (LSB, the last binary digit in the right-most place) stands for a 1, the next one to the left stands for a 2, the next one stands for a 4, and the last one and "most significant bit" (MSB, or bit at the left-most place) stands for an 8. This way, 1001_2 is $8 + 1 = 9$, $1100_2 = 8 + 4 = 12$, $0110_2 = 4 + 2 = 6$, and so forth. Learning these conversions now will save you time later. You can verify that you can easily convert the first 16 rows of Table 2.1 using this method.

Hexadecimal to Binary and Binary to Hexadecimal   You can easily convert hexadecimal digits to binary because the 16 characters that make up the digits in this base system have a corresponding unique combination of 4 binary bits. Conversely, all the possible combinations of 4 bits have a corresponding unique hexadecimal digit (see Table 2.1). For example, 10100111_2 can be divided into two groups of four digits: 1010_2 and 0111_2. Each digit can be converted individually: $1010_2 = 10 = A_{16}$ and $0111_2 = 7$. Therefore, $1010\ 0111_2 = A7_{16}$. The reverse is also true. For example, the number $15C_{16}$ gets converted to binary by converting each digit separately: $15C_{16} = 000101011100_2$.

■ **EXERCISE 2.3**

Convert B3F$_{16}$ *to binary.*

■ **EXERCISE 2.4**

Convert 10110110_2 *to hexadecimal.*

Decimal to Binary or Another Base r   To convert a decimal number to binary you divide the number consecutively by two. The remainders in the divisions are the binary digits. For example, 82_{10} gets converted the following way: $\frac{82}{2} = 41 + \frac{0}{2}$, $\frac{41}{2} = 20 + \frac{1}{2}$, $\frac{20}{2} = 10 + \frac{0}{2}$, $\frac{10}{2} = 5 + \frac{0}{2}$, $\frac{5}{2} = 2 + \frac{1}{2}$, $\frac{2}{2} = 1 + \frac{0}{2}$, and, finally, $\frac{1}{2} = 0 + \frac{1}{2}$. The first remainder is the least significant bit (LSB), and the last remainder is the most significant bit (MSB). Thus, $82_{10} = 1010010_2 = 52_{16}$. This operation is displayed here:

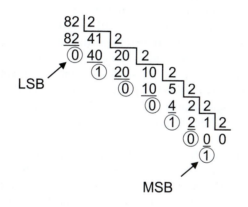

Conversion from decimal to base r involves the same procedure: consecutive divisions by r, with the remainders becoming the digits of the number in base r.

2.1.2 Arithmetic Operations

Addition   When you add two digits in base r, some results give you a number N less than r. In this case, N is another one-digit number in base r. If the result of the addition is greater than, or equal to r, then you generate a "carry," a two-digit number consisting of a 1 followed by the digit $N - r$. In binary, we have four combinations: $0_2 + 0_2 = 0_2, 0_2 + 1_2 = 1_2$, $1_2 + 0_2 = 1_2$, and $1_2 + 1_2 = 10_2$. The last operation is the one that generates a carry. If you add two numbers that have more than one digit, the same rule applies, digit by digit; the only difference is that you may get an incoming carry from the previous digit operation. This way, you could face $1_2 + 1_2 + 1_2 = 11_2$, where one of the 1's is an incoming carry.

Subtraction   In subtraction you operate under the same rules as in ordinary arithmetic. If the one-digit minuend in base r is greater than or equal to the subtrahend, then the answer is a one-digit number in base r. If the reverse is true, then a "borrow" is generated: The minuend then borrows one unit from the next-higher digit and the answer is (minuend) + r − (subtrahend). The borrow leaves the next-higher digit of the minuend with one unit less.

■ **EXERCISE 2.5**

Calculate the sum and difference of 101101_2 *and* 001110_2.

2.2 CODES

A computer stores all information in binary. All the arithmetic and logic operations in a computer are done in binary. Executable files contain a long sequence of binary digits that tell the computer what to do. These files are therefore referred to as binary files. For example, files with extensions ".exe" are executable binary files (in PCs). However, the computer does not do pure computation with all files. Many files are used for data storage purposes only. In the same way, we also need to store text. Computers use two types of binary codes to represent numbers and characters: Numeric and alphanumeric. A numeric code is a way to represent a number in the computer for storage instead of computational purposes. A popular numeric code is the binary-coded decimal (BCD). It consists of converting a decimal number to binary, digit by digit. For example, $46 = 0100_2$ 0110_2 in BCD. The true binary conversion of 46 is $0010\ 1110_2$, or $2E_{16}$. Converting a decimal number to BCD is faster than converting it to its true binary equivalent.

The most widely used system for representing characters is the ASCII code (American Standard Code for Information Interchange). It provides a number representation for many common characters and controls involved in text processing. Each character is represented by seven binary digits (or an 8-bit number with its MSB always zero). Table 2.2 gives the ASCII codes in hexadecimal. You can see codes for all the characters that appear in a standard U.S.-language keyboard.

In addition, other codes represent text-processing actions such as carriage return (CR), line feed (LF), backspace (BS), and beep (BEL). These were originally developed for teletype transmissions. The ASCII code has 7 bits, originally designed to minimize the number of bits required to send characters in messages. Over the years, it has evolved

Table 2.2. Table of ASCII Codes

00 NUL	10 DLE	20 SP	30 0	40 @	50 P	60 `	70 p
01 SOH	11 DC1	21 !	31 1	41 A	51 Q	61 a	71 q
02 STX	12 DC2	22 "	32 2	42 B	52 R	62 b	72 r
03 ETX	13 DC3	23 #	33 3	43 C	53 S	63 c	73 s
04 EOT	14 DC4	24 $	34 4	44 D	54 T	64 d	74 t
05 ENQ	15 NAK	25 %	35 5	45 E	55 U	65 e	75 u
06 ACK	16 SYN	26 &	36 6	46 F	56 V	66 f	76 v
07 BEL	17 ETB	27 '	37 7	47 G	57 W	67 g	77 w
08 BS	18 CAN	28 (	38 8	48 H	58 X	68 h	78 x
09 HT	19 EM	29)	39 9	49 I	59 Y	69 i	79 y
0A LF	1A SUB	2A *	3A :	4A J	5A Z	6A j	7A z
0B VT	1B ESC	2B +	3B ;	4B K	5B [	6B k	7B {
0C FF	1C FS	2C ,	3C ¡	4C L	5C \	6C l	7C |
0D CR	1D GS	2D –	3D =	4D M	5D]	6D m	7D }
0E SO	1E RS	2E .	3E ¿	4E N	5E ^	6E n	7E ~
0F SI	1F US	2F /	3F ?	4F O	5F _	6F o	7F DEL

into Unicode, to include many other characters for different languages. The UTF-8 code, a variation of Unicode, uses 1 or more bytes to encode a large set of characters for use in modern operating systems and in the World Wide Web.

■ **EXERCISE 2.6**

What character does the ASCII code 0101 1101 *represent?*

2.3   SIGNED NUMBERS

How are negative numbers represented in binary? Ideally, we want to represent them the same way we do in standard arithmetic: with a minus sign in front of the number. The closest that we can get to this is to have the MSB of a binary number be the sign bit: 0 for positive and 1 for negative. All representations of negative binary numbers do that in some way.

Sign and Magnitude   In this representation, the MSB is exclusively the sign bit independently of the actual number, as shown in Table 2.3. However, although this convention is convenient for us to identify, it is not efficient from the computer's point of view; it makes computations awkward and time consuming. In two other signed representations, the MSB is an integral part of the number. This allows the computer to do arithmetic in a more straightforward way. These representations are described next.

Table 2.3.   Different Ways to Express Negative Numbers in Binary Form

Number	SAM	One's complement	Two's complement	Unsigned
00000000	0	0	0	0
00000001	+1	+1	+1	1
⋮	⋮	⋮	⋮	⋮
01111110	+126	+126	+126	126
01111111	+127	+127	+127	127
10000000	−0	−127	−128	128
10000001	−1	−126	−127	129
10000010	−2	−125	−126	130
⋮	⋮	⋮	⋮	⋮
11111110	−126	−1	−2	254
11111111	−127	−0	−1	255

One's Complement   In one's complement, the negative number of an n-digit number is

$$\overline{N}_{1c} = (2^n - 1) - N \tag{2.3}$$

For example, for a number of four digits

$$-0011_2 = \overline{0011}_2 = 1111_2 - 0011_2 = 1100_2$$

A practical way to do this conversion is to replace all the zeros with ones, and replace all ones with zeros.

Two's Complement   In this representation the negative number is

$$\overline{N}_{2c} = 2^n - N \tag{2.4}$$

The two's complement of the previous example is $\overline{0011}_2 = 10000_2 - 0011_2 = 1101_2$. Notice that $\overline{N}_{2c} = \overline{N}_{1c} + 1$. Thus, a practical way to get the two's complement is to get the one's complement (replace 0's with 1's and vice versa) and then add a 1 to it. Two's complement is good for computation because, to subtract two numbers, you just add the minuend to the 2's complement of the subtrahend. For example

$$0101_2 - 0010_2 = 0101_2 + \overline{0010}_2 = 0101_2 + 1110_2 = 10011_2 = 0011_2$$

We ignore the carry generated in the last operation. In both one's and two's complement, the MSB also serves as an identifier of the sign. That is, if the MSB of a number is 0, it is a positive number, and if the MSB is 1, then the number is negative.

If we have 8-bit numbers, the highest unsigned value is $2^8 - 1 = 255 = FF_{16}$, and the highest signed number is $2^7 - 1 = 127 = 7F_{16}$. The cost of carrying the sign is to lose the maximum value of the numbers (see Table 2.3).

2.4   BINARY FUNCTIONS

Complex binary functions can be put in terms of a set of simple functions. We can represent these simple functions by a mathematical/logic expression, a logic symbol, or a "truth table." The truth table is a list of the function's output values for all the possible combinations of the inputs.

2.4.1   Fundamental Gates

We need to define a few basic binary functions. These are also known as gates. The first three gates, OR, AND, and NOT, are the fundamental operations of binary logic. With them, you can represent any binary function.

OR Gate: $F = A + B$ Table 2.4 gives the truth table for the OR operation. The output is 1 when any of the inputs is 1. The table also gives the electronic symbol for the two-input OR gate. This logic operation can take more than two inputs. In all cases, the output is 1 when any of the inputs is 1, and 0 when all the inputs are 0.

Table 2.4.   Truth Table and Electronic Symbol for the OR Gate

A	B	F	Symbol
0	0	0	
0	1	1	
1	0	1	
1	1	1	

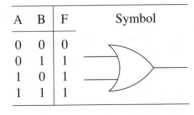

NOTATION ALERT: From here on, symbol "+" represents the OR operation. When we want to represent addition, we spell it out as "plus." The symbol for the electronic two-input OR gate is shown next.

Table 2.5 lists fundamental-gate ICs. Digital ICs are coded as 74XXYY, where XX corresponds to the electronic technology (LS or HC, discussed in Section 2.5), and YY is the type of gate listed in the first column of Table 2.5. Also check the web site www.datasheetcatalog.com for a comprehensive list of gate data sheets. The table also shows the electronic type-number for OR gates with more than two inputs: 3-input and 4-input gates. If you need an OR gate with more than 4 inputs, you can use several OR gates, and input their output to another OR gate. For example if want to OR six variables, A, B, C, D, E, and F, we can do

$$A+B+C+D+E+F = (A+B)+(C+D)+(E+F) = (A+B+C)+(D+E+F) \quad (2.5)$$

The parenthesis represent 2-input (middle term) and 3-input (last term) OR gates.

Table 2.5. Useful Fundamental Gate ICs

IC Type #	Description
32	2–input OR
4075[†]	3–input OR
4072[†]	4–input OR
08	2–input AND
11	3–input AND
21	4–input AND
04	NOT
14	NOT with Schmitt trigger[‡]

[†]HC only.

[‡]$0 \rightarrow 1$ threshold is 1.6 V; $1 \rightarrow 0$ threshold is 0.8 V.

AND Gate: F = A · B   The AND operation is defined by the truth table of Table 2.6. The output is 1 only when all inputs are 1. The symbol "·" is used here to represent the AND operation. The electronic symbol for the 2-input AND gate is shown in Table 2.6.

Table 2.6. Truth Table and Electronic Symbol for the AND Gate

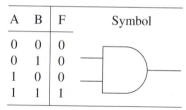

A	B	F	Symbol
0	0	0	
0	1	0	
1	0	0	
1	1	1	

Because in logic there is no arithmetic multiplication, only the AND operation, we will represent the AND operation the same way we do in arithmetic. That is, we represent the AND operation between variables A and B with either $A \cdot B$, or AB.

Similarly to the OR gate, an AND gate can take more than two inputs. The output of the multiple-input AND gate will be 1 when all the inputs are 1, and 0 otherwise. Table 2.5 gives the electronic IC denomination for 2-, 3-, and 4-input AND gates. If we want an AND gate with more than 4 inputs, we can concatenate AND gates the same way we did with OR gates:

$$ABCDEF = (AB)(CD)(EF) = (ABC)(DEF) \quad (2.6)$$

The parenthesis represent 2-input (middle term) and 3-input (last term) AND gates.

NOT Gate: F = A'   The NOT operation provides the inverse of the input variable, as defined in Table 2.7.

Table 2.7.   Truth Table and Electronic Symbol for the NOT Gate

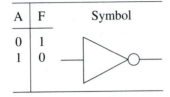

A	F	Symbol
0	1	
1	0	

We use two symbols interchangeably to represent this operation: the variable primed, A', or a line over the variable, $\overline{A}$. The electronic symbol is shown in Table 2.7.

2.4.2   Universal Gates

The NAND and NOR gates are equivalent to AND and OR gates followed by a NOT gate, respectively. They are important because they are universal. That is, an arbitrary binary function can be implemented by a circuit made up entirely of gates of the same kind. Mathematically, these gates are tricky because they do not obey the associative property.

NAND Gate: F = (A · B)'   This is the most popular gate and, electronically, the simplest one. It is defined by the truth table and symbol of Table 2.8. The output is a 1 when any of the inputs is a 0. It is the complement of the AND operation:

$$F = A \uparrow B = (A \cdot B)' \tag{2.7}$$

Table 2.8.   Truth Table and Electronic Symbol for the NAND Gate

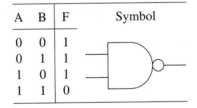

A	B	F	Symbol
0	0	1	
0	1	1	
1	0	1	
1	1	0	

Table 2.9 shows NAND gates with multiple inputs. It is important to note that NAND gates do not obey the associative property as OR and AND gates do. That is,

$$A \uparrow B \uparrow C \neq (A \uparrow B) \uparrow C \tag{2.8}$$

or

$$A \uparrow B \uparrow C \uparrow D \neq (A \uparrow B) \uparrow (C \uparrow D) \tag{2.9}$$

The fundamental gates can be constructed with NAND gates as shown in Figure 2.1. The next chapter describes the rationale behind those implementations.

■ **EXERCISE 2.7**

Verify the relationship between the OR gate and NAND gates of Figure 2.1 using the truth table.

Table 2.9.   Useful Universal Gate ICs

IC Type #	Description
00	2–input NAND
10	3–input NAND
20	4–input NAND
30	8–input NAND
133	13–input NAND
132	2–input NAND with Schmitt trigger
13	4–input NAND with Schmitt trigger
02	2–input NOR
27	3–input NOR
25	4–input NOR

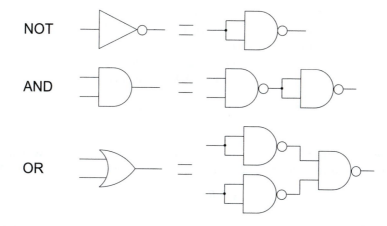

Figure 2.1.   The function of the three fundamental gates can be implemented with NAND gates.

NOR Gate: *F = (A + B)'* The NOR gate operates as follows: The output is a 1 only when all the inputs are 0. It is the complement of the OR gate:

$$F = A \downarrow B = (A + B)' \tag{2.10}$$

Its truth table is shown in Table 2.10. Its electronic symbol follows the same convention as the NAND gate: an OR gate followed by a circle to represent the complement.

2.4.3 Specialty Gates

The next two gates have specific uses. The XOR and XNOR gates are used for making adders and parity generators. These are functions that we will cover soon. Individual ICs are listed in Table 2.11. In recent years, the XOR gate has received much attention for doing reversible quantum computation. It is the basic building block of the CNOT gate.

Table 2.10. Truth Table and Electronic Symbol for the NOR Gate

A	B	F	Symbol
0	0	1	
0	1	0	
1	0	0	
1	1	0	

Table 2.11. Useful Specialty and Buffer ICs

IC Type #	Description
86	2–input XOR
266[†]	2–input XNOR
30	4–input NAND
365[‡]	Buffer
125,[‡] 126[‡]	Quad buffer
241,[‡] 244[‡]	Octal buffer/line driver
243,[‡] 8216[#]	Quad bus transceiver
245	Octal bus transceiver
13	4–input NAND with Schmitt trigger

[†]Open-collector/drain, [‡]Tristate output, [#]CMOS only

XOR Gate: $F = A \oplus B$ The XOR function of two inputs is 1 when the two inputs are not the same, as shown in the truth table of Table 2.12. It is represented by the function

$$F = A \oplus B = AB' + A'B \tag{2.11}$$

Table 2.12. Truth Table and Electronic Symbol for the XOR Gate

A	B	F	Symbol
0	0	0	
0	1	1	
1	0	1	
1	1	0	

You can figure this gate the following way: When input $x = 0$ the output is $F = y$, and when $x = 1$ the output is $F = y'$. That is, the output F is B or $\overline{B}$ depending on the condition A.

You can use 2-input XOR gates to construct an exclusive OR of more than two inputs. This is because XOR obeys the associative property:

$$(A \oplus B) \oplus C = A \oplus (B \oplus C)$$

We cannot do this with NAND and NOR gates. One can make an XOR of more variables by putting them in tandem, as mentioned earlier. The output of such a circuit is useful for parity checking because the output is a 1 when an odd number of inputs is a 1.

XNOR Gate: F = A ⊙ B   The XNOR function is also known as *equivalence*. The output is 1 when the two inputs are the same, as defined by the truth table and symbol of Table 2.13.

Table 2.13. Truth Table and Electronic Symbol for the XNOR Gate

A	B	F	Symbol
0	0	1	
0	1	0	
1	0	0	
1	1	1	

It is represented by the function:

$$F = A \odot B = AB + A'B' \tag{2.12}$$

As shown in Chapter 4, this function is useful for comparing binary numbers.

2.4.4 Utilitarian Gates

These are gates that respond to more practical aspects of circuit design.

BUFFER Gate: F = A   This gate provides the trivial identity function. We use it for isolating valuable circuitry. That is, at the input and output of an "important" device, we have buffers connected to the outside world. If the external device fails in some way, exposing the input or output to a dangerous electrical surge, the buffer is the one that gets destroyed and thus protects the valuable circuitry. Otherwise, the entire circuitry may be compromised. The truth table and symbol for the buffer are given in Table 2.14.

Table 2.14. Truth Table and Electronic Symbol for the Buffer Gate

A	F	Symbol
0	0	
1	1	

Tristate Buffer   Many devices require complete electrical isolation when not utilized. That is, we need to be able to disable outputs so that when they are not in use, they have no electrical output whatsoever. The "tristate" buffer provides this function. The truth table and symbol for a tristate buffer are shown in Table 2.15. A third input, T, puts the output in a "high impedance" (H.I.) state when false. In this state, the output of the gate is effectively a nonconnection.

This device is useful for connecting the output of several devices to a common line. The last gate before the common line is a tristate buffer. The logic is then arranged through a "decoder" (patience, next chapter) that enables only one of the T inputs of the devices.

Table 2.15.   Truth Table and Electronic Symbol for the Tristate-Buffer Gate

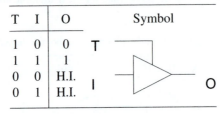

T	I	O	Symbol
1	0	0	
1	1	1	
0	0	H.I.	
0	1	H.I.	

2.4.5 Matrix Representation

The use of matrices for binary operations is going through a revival because of the interest in quantum computation. In this section, we explore this method, which will become necessary when a quantum computer is available. So before our text becomes obsolete, we take a peek at what will make it obsolete. If this section seems to depart too much from the discussion, you can skip it without any consequences.

One warning: We will introduce new symbols to make some distinctions. For example, to denote a variable and distinguish it from a number, we put it within a "ket." Physicist Paul Dirac, one of the founding fathers of quantum mechanics, introduced this notation to simplify quantum mechanics algebra. However, we will not be giving any quantum meaning to our kets. The formalism uses linear algebra for the operations.

In this representation, the bits are represented by column vectors:

$$|0\rangle = \begin{pmatrix} 1 \\ 0 \end{pmatrix} \quad |1\rangle = \begin{pmatrix} 0 \\ 1 \end{pmatrix}$$

A gate operating on 1 bit is represented by a 2×2 matrix. The matrix for the buffer gate is the identity matrix:

$$I = \begin{pmatrix} 1 & 0 \\ 0 & 1 \end{pmatrix} \tag{2.13}$$

so that

$$\begin{pmatrix} 1 & 0 \\ 0 & 1 \end{pmatrix} \begin{pmatrix} 1 \\ 0 \end{pmatrix} = \begin{pmatrix} 1 \\ 0 \end{pmatrix}$$

The NOT gate flips the bit. It is given by

$$\text{NOT} = \begin{pmatrix} 0 & 1 \\ 1 & 0 \end{pmatrix} \tag{2.14}$$

We can easily verify that the NOT matrix indeed flips the bit:

$$\begin{pmatrix} 0 & 1 \\ 1 & 0 \end{pmatrix} \begin{pmatrix} 1 \\ 0 \end{pmatrix} = \begin{pmatrix} 0 \\ 1 \end{pmatrix}$$

The size of the vector for an n-input gate is 2^n, where each vector row corresponds to a combination of the input. For 1-bit operations, we have two-dimensional vectors. 2-bit gates that have a $2 \times 2 = 4$ possible combination of inputs are four-dimensional. That way:

$$|00\rangle = \begin{pmatrix} 1 \\ 0 \\ 0 \\ 0 \end{pmatrix} \quad |01\rangle = \begin{pmatrix} 0 \\ 1 \\ 0 \\ 0 \end{pmatrix} \quad |10\rangle = \begin{pmatrix} 0 \\ 0 \\ 1 \\ 0 \end{pmatrix} \quad |11\rangle = \begin{pmatrix} 0 \\ 0 \\ 0 \\ 1 \end{pmatrix} \tag{2.15}$$

A 2-input gate has a single output, so its matrix should be a 2×4. An OR gate is given by

$$\text{OR} = \begin{pmatrix} 1 & 0 & 0 & 0 \\ 0 & 1 & 1 & 1 \end{pmatrix} \tag{2.16}$$

One can easily see that $(\text{OR})|00\rangle = |0\rangle$ or $(\text{OR})|01\rangle = |1\rangle$.

■ **EXERCISE 2.8**

Find the matrix for the AND gate.

Suppose that we want to create the matrix of a NOR gate indirectly, by using an OR matrix and a NOT matrix. This involves two sequential operations. Suppose that the 2-bit input is $|AB\rangle$. After going through the OR gate it will be a single bit: $|C\rangle = (\text{OR})|AB\rangle$. After the OR gate, we operate the NOT gate, obtaining

$$|D\rangle = (\text{NOT})|C\rangle = (\text{NOT})(\text{OR})|AB\rangle$$

That is, $(\text{NOR}) = (\text{NOT})(\text{OR})$, a multiplication of the two matrices. Note that the order of the operations, from right to left, is important.

As you can see, matrices can represent all gate operations. Consider one final point. How do we account for the case when two 1-bit variables have single-bit operations before they go on into a 2-bit operation? Suppose that variables $|A\rangle$ and $|B\rangle$ go through NOT gates before going into an AND gate. How do we represent this by a single matrix? In this case, we combine variables $|A\rangle$ and $|B\rangle$ into a single 2-bit variable $|AB\rangle$. The proper ordering of the bits of the 1-bit vectors to form the 2-bit vector follows the *tensor product*:

$$|AB\rangle = |A\rangle \otimes |B\rangle \tag{2.17}$$

More specifically, this means that if $|A\rangle = \begin{pmatrix} a_1 \\ a_2 \end{pmatrix}$ and if $|B\rangle = \begin{pmatrix} b_1 \\ b_2 \end{pmatrix}$, then

$$|AB\rangle = \begin{pmatrix} a_1 b_1 \\ a_1 b_2 \\ a_2 b_1 \\ a_2 b_2 \end{pmatrix}$$

Note that the ordering in equation 2.15 follows the same rule. Following the same logic, the matrix that should operate on $|AB\rangle$ should be the tensor product of the matrices that operate the individual bits $|A\rangle$ and $|B\rangle$. The tensor product of two 2×2 matrices is a 4×4 matrix. The general rule is this: The tensor product of

$$(\text{M}) = \begin{pmatrix} m_{11} & m_{12} \\ m_{21} & m_{22} \end{pmatrix}$$

and

$$(\text{N}) = \begin{pmatrix} n_{11} & n_{12} \\ n_{21} & n_{22} \end{pmatrix}$$

is this:

$$(\text{M}) \otimes (\text{N}) = \begin{pmatrix} m_{11}n_{11} & m_{11}n_{12} & m_{12}n_{11} & m_{12}n_{12} \\ m_{11}n_{21} & m_{11}n_{22} & m_{12}n_{21} & m_{12}n_{22} \\ m_{21}n_{11} & m_{21}n_{12} & m_{22}n_{11} & m_{22}n_{12} \\ m_{21}n_{21} & m_{21}n_{22} & m_{22}n_{21} & m_{22}n_{22} \end{pmatrix} \tag{2.18}$$

Following the same rule, then, (NOT) $\otimes$ (NOT) is this:

$$\begin{pmatrix} 0 & 1 \\ 1 & 0 \end{pmatrix} \otimes \begin{pmatrix} 0 & 1 \\ 1 & 0 \end{pmatrix} = \left(\begin{array}{cc|cc} 0\cdot0 & 0\cdot1 & 1\cdot0 & 1\cdot1 \\ 0\cdot1 & 0\cdot0 & 1\cdot1 & 1\cdot0 \\ \hline 1\cdot0 & 1\cdot1 & 0\cdot0 & 0\cdot1 \\ 1\cdot1 & 1\cdot0 & 0\cdot1 & 0\cdot0 \end{array}\right) = \begin{pmatrix} 0 & 0 & 0 & 1 \\ 0 & 0 & 1 & 0 \\ 0 & 1 & 0 & 0 \\ 1 & 0 & 0 & 0 \end{pmatrix}$$

The final operation will then be (AND)[(NOT) $\otimes$ (NOT)].

■ **EXERCISE 2.9**

Suppose we have a gate G where one input, $|A\rangle$, goes through a NOT gate and then into an AND gate, and the other input, $|B\rangle$, goes straight into the AND gate.

1. *Find the matrix for G.*
2. *Fill in the truth table for gate G by calculating its output using matrix operations.*

2.5   LOGIC FAMILIES

Several families of electronic gates have emerged over the years. These are transistor-transistor logic (TTL), complementary metal-oxide semiconductor (CMOS), and emitter-coupled logic (ECL). The overriding performance parameters that have driven these technologies to improvement are speed and power consumption. The speed is normally represented by the time it takes for a signal to go through a gate, also known as the propagation delay t_{pd}. That is, the effect of a transition from, say, 0 to 1 in the input of a gate appears at the output after a delay t_{pd}. The faster the gate, the lower t_{pd}. An input signal should not be changing from 0 to 1 or from 1 to 0 in a time shorter than t_{pd}. Otherwise, the gate would not be able to keep up with the changing input. Therefore, t_{pd} sets a limit on the highest frequency that can be achieved ($f_{max} \sim 1/t_{pd}$). Many families of digital gates exist within the respective categories of TTL, CMOS, and ECL. The families represent the evolution of the respective technologies.

The TTL technology evolved from plain TTL (such as 7400 for a NAND gate; 1960s), to Schottky, "S" (as in 74S00), which is faster but not low power (1970s); to low-power Schottky, "LS" (as in 74LS00; 1970s); to advanced low-power Schottky, "ALS" (as in 74ALS00; 1980s); and fast, "F" (as in 74F00; 1980s). Bipolar transistors are at the core of this technology. However, they consume considerable power because transistors are driven by currents. They are rugged and versatile, and can be used widely in today's applications.

The CMOS technology evolved from the old MOS (metal oxide semiconductor) technology. These devices are driven by voltages, so they consume little power. They were initially slow, but they, too, evolved into faster devices. The introduction of complementary MOS devices revolutionized this technology. The initial series were "C" (as in 74C00), but they are now obsolete. They were followed by the high-speed CMOS, "HC" and "HCT" (as in 74HC00 and 74HCT00; 1980s), and more recently, advanced high speed CMOS, "AHC" and "AHCT" (as in 74AHC00 and 74AHCT00), used widely in the semiconductor industry. The best option to use in the lab in circuits with discrete components is still TTL's LS. When not available, the HCT family is second best because it interfaces well with TTL.

The low-power consumption of CMOS makes this family a suitable choice for battery-operated devices. The latest trend is to go to lower voltages, with a number of families

Table 2.16. Specifications of NAND Gates Made with Different Technologies

Family	t_{pd} (ns)	f_{max} (MHz)	P_{diss} (mW/gate)
TTL-LS	10	45	2
TTL-ALS	5	80	2
TTL-F	3.5	100	5.4
CMOS-HC/HCT	3	125	10^{-4}

now available, "LV" and other, such as "LVC" and "ALVC" (as in 74LV00 or 74ALVC00; 1990s). This enables the use of low-voltage batteries as the power source.

Table 2.16 lists these parameters for a range of versions of each family. The lab for this course can use both CMOS and TTL gates. The family of choice is still TTL because TTL gates can put up with a lot of abuse. However, some new ICs exist only in HC or HCT. In those cases, we can mix the two families. An additional parameter of logical gates is "fanout." This is the maximum number of gates that a given gate can drive. For the technologies shown previously, it is about 20.

In addition, because of variations in performance from gate to gate, gates' inputs and outputs are specified to comply with standard tolerances. Tables 2.17 and 2.18 list the tolerances for TTL-LS and CMOS-HC families, respectively. They assume *positive logic*, or a high voltage for a logic 1 and a low voltage for a logic 0. We can invert the two in some applications. This is also known as *negative logic*. It is tricky, though, because the gate definitions change. ICs follow positive logic. For example, the gates in an AND-gate IC work as AND gates in positive logic but as OR gates in negative logic.

In TTL, a gate takes any input voltage between 2 V and 5.5 V as a logic 1, whereas CMOS-HC has a more restricted range: between 3.5 V and 5 V. Similarly, the input range for logic 0 is −0.5 V to 0.8 V for TTL, and 0 V to 1.5 V for CMOS-HC. However, if the input is outside the range for a given family (for example, between 0.8 V and 2 V for TTL), then there is no way to predict which logic value the gate will adopt.

Figure 2.2 shows a comparison of the tolerances between TTL and CMOS. As we can see in Tables 2.17 and 2.18, the CMOS-HC gates have lower tolerances. For this reason, CMOS-HC gates can drive TTL gates, but not the other way around. A solution to this is to use the CMOS-HCT family of gates, which have the same input tolerances as TTL but the same output tolerances as HC-CMOS. The CMOS-HCT is a comprehensive family of gates dedicated at the interfacing of TTL to CMOS. By using an HCT gate after a TTL gate, we can drive CMOS with a TTL gate. HC gates can drive TTL gates with no problem. Table 2.19 summarizes the possible gate connections.

Table 2.17. TTL-LS Input/Output Characteristics

	Value	V_{lower}	V_{higher}
Input	0	−0.5	0.8
Input	1	2	5.5
Output	0	0	0.4
Output	1	2.5	5.0

Table 2.18. CMOS-HC Input/Output Characteristics

	Value	V_{lower}	V_{higher}
Input	0	0	1.5
Input	1	3.5	5
Output	0	0	0.05
Output	1	4.95	5.0

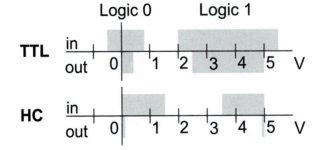

Figure 2.2. Comparison of input/output tolerances of TTL-LS and CMOS-HC

Table 2.19. Possible Connections Between HC, HCT and TTL Digital Families

Drive Gate	Driven Gate	Correct?
HC	HCT	Yes
HCT	HC	Yes
HCT	TTL	Yes
TTL	HCT	Yes
HC	TTL	Yes
TTL	HC	No

In addition, the "low-voltage" series of gates LV, LVC, and ALVC operate at low supply voltages, between 1 and 5 V. The input and output logic levels scale with the supply voltage. Table 2.20 gives a few examples. A disadvantage of the LV technology is that most ICs come only with surface mount connections, so they are not easily wired to breadboards.

2.6 IC WIRINGS

Starting with this chapter, we will cover wiring logic circuits. Suppose that we want to wire the circuit of Figure 2.3. You see two NAND gates that are being used as NOT gates to implement a buffer. You also see creative ways to implement a NOT gate with a NAND gate. Is something else missing? The figure does not show the power connections. This is common in digital circuits. It avoids cluttering the diagram with trivial power connections. The circuit assumes that you did the obvious: connect the power. It is your responsibility to worry about hooking up the power when you wire the circuit.

Table 2.20.   LV/LVC/ALVC Input/Output Characteristics for Logic 1

V_{CC}	$V_{input-min}$	V_{output}
1.2	0.9	1.2
2.0	1.4	2.0
2.7-3.6	2.0	V_{CC}
4.5-5.0[†]	0.7 V_{CC}	V_{CC}

[†]LV only.

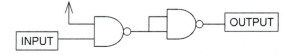

Figure 2.3.   A digital circuit.

If you look up the data sheet for the IC that holds NAND gates, 74HC00, you will find that each IC holds four NAND gates, with power and ground connections that are common to all four gates. Figure 2.4 shows the pin connections for the 74HC00. In most ICs, the numbering of pins follows a counter-clockwise ordering, with the starting pin, 1, in the upper left. The upper side of the IC is labeled with a notch, as shown. The diagram also shows the power and ground connections, which are almost always upper right (pin 14) for the positive side of the supply and lower left (pin 7) for ground. Again, the negative side of the supply is the common line or signal ground that is normally connected to the ground connection of the external devices. In the figure, we have drawn a possible connection that implements the circuit.

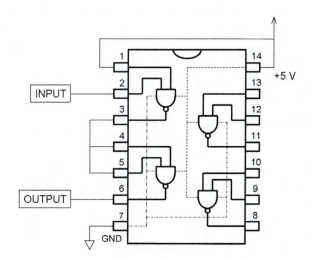

Figure 2.4.   Diagram of the actual wiring of the circuit of Figure 2.3.

2.7 PROBLEMS

1. Explain the difference between analog and digital signals.

2. Convert the following binary numbers to decimals:

    (a) 11101

    (b) 10111.11

3. Ancient Mayans, prolific astronomers and time keepers who lived in Yucatan around 1000 AD (and who did *not* predict the end of the world in 2012), used the vigesimal number system (base 20). For calendars, they modified the number system slightly so that the digits would be close to multiples of a year (360 days). The meaning of a four-digit number in a Mayan calendar is:

$$A_3 \cdot A_2 \cdot A_1 \cdot A_0 = A_3 \times 18 \times 20^2 + A_2 \times 18 \times 20 + A_1 \times 20 + A_0$$

    Here, all the digits except for A_1 can range from 0 to 19, but A_1 can range only from 0 to 17. For example, the number 1.0.0 is 360 days (we separate here the digits with a period), and 1.0.2.5 is $7,200 + 40 + 5 = 7,245$. The Mayans had 20 characters, including zero. These were elaborate, mean-looking anthropomorphic figures, but apparently also used a simpler system of three characters and positions: a bar $= 5$, a dot $= 1$, and a shell $= 0$. Each "digit" was a grouped combination of bars and dots, or a shell. The positions were oriented vertically, with the least significant position lowest and most significant position the highest. The lowest position (least significant place) was the units, the second position had the 20s, the third position were the 360s (18×20), the fourth position were the 7,200s (18×400), and so on. Figure 2.5 is a reproduction of a section of the Eclipse table of a Mayan document (the Dresden Codex) used to predict eclipses.[1] It gives several groupings of numbers denoting partial and cumulative days between lunar months. We have selected several numbers. For example number A is read (vertically) as 1.6.0.0 or (dot $= 1) \times 7,200 + $ (dot plus bar $= 6) \times 360 = 9,360$.

    (a) What are the decimal equivalents of the numbers labeled B, C, and E?

    (b) It is trivial to see that $A + B = C$ (it is a cumulative table, so the top numbers are the totals and the bottom ones are the increments) and that $B = D$. Show $C + D = E$. Do the addition within the Mayan number system (do not convert to decimal to do the addition).

    (c) Deduce the value F. It represents the predicted number of days between two eclipses. Explain your reasoning and calculation.

4. The IP addresses used for Internet networks consist of four 8-bit binary numbers separated by periods. A valid IP is, for example, 149.43.164.45, where the displayed numbers are in decimal. In Internet networks, the MSB of the left-most number has a special meaning. If it is a 0, the IP address is from a Class A (large) network like one of a major international company or a government. If the MSB is 1 and the binary digit next to it is a 0, then it is a Class B (medium) network, like one of a university campus. If the binary number starts with 110, then it is a Class C (small) network, such as one of a small business. Convert the IP address given earlier to hexadecimal and state what type of network it corresponds to.

[1]For more information see A.F. Aveni, in *Stairways to the Stars* (New York: Wiley, 1997), 105ff.

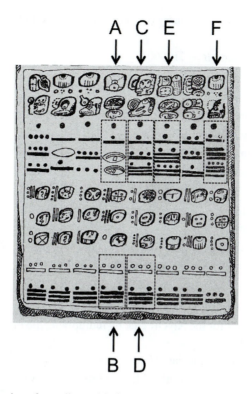

Figure 2.5.   Reproduction of an eclipse table from the Dresden Codex, an ancient document made by the Mayans, which reveals their unique number system (see Problem 3). Taken with permission from J.A. Villacorta and C.A. Villacorta in *Codices Mayas* (Guatemala C.A., 1977).

5. For the number 45, give the following information:

   (a) Convert it to binary.

   (b) Based on the previous part, convert the number to hexadecimal.

   (c) Give its BCD representation in a BCD file.

   (d) If the two-digit number were to be stored in an ASCII (text) file as character 4 followed by character 5, what ASCII codes would represent those characters?

   (e) If 45_{16} were an ASCII code, what character it would represent?

6. Assuming that we are working with 8-bit numbers, consider the number $N = 10101101_2$.

   (a) Give its one's complement.

   (b) Give its two's complement.

   (c) Is N a positive or negative number? Why?

7. Give the negative of 57_{16} in the following:

   (a) Sign and magnitude

   (b) One's complement

   (c) Two's complement

8. Subtract 47_{16} from 66_{16} in binary by adding the two's complement of the subtrahend to the minuend.

9. If $A = 01101101_2$ and $B = 11000011_2$

   (a) Calculate A *plus* B in binary.

   (b) Calculate B *minus* A in binary.

   (c) Check the previous result by adding B to the two's complement of A.

   (d) Convert A to hexadecimal.

   (e) Convert B to hexadecimal.

   (f) Without a calculator, convert A to decimal.

   (g) Without a calculator, convert B to decimal.

10. Use truth tables to help you construct NOT, 2-input OR and 2-input AND gates using purely NOR gates.

11. Use truth tables to show that the *complement* of the XOR of three variables, $(A \oplus B) \oplus C$, is a 1 only when an even number of inputs is 1.

12. Do the following gate-matrix operations:

    (a) Find the matrix for the XOR gate.

    (b) Find the matrix for the XNOR gate by multiplying the matrices of the XOR gate and the NOT gate.

    (c) Find the matrix for the gate $G = (A'B')'$.

    (d) Find the entries for the truth table for G using matrix operations.

13. In the partial circuit diagram of Figure 2.6, two different digital lines must be connected to a common line. Tristate gates help us avoid conflicts where one line puts a logic 0 and another puts a logic 1. We then define a protocol where a third-line ENABLE controls traffic. When ENABLE $= 1$, it allows the first input to go through and disables the second one; conversely, when ENABLE $= 0$, it allows the second one to go through, disabling the first one. Using any gates that you deem necessary, complete the circuit so that it enables the desired protocol.

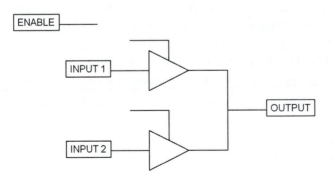

Figure 2.6.   Circuit for Problem 13.

14. In the circuit of Figure 2.7, we want to implement a bi-directional transceiver. The idea is that two devices need to communicate with each other over a single line. Each

device has input and output capabilities. For simplicity, we call the devices "left" and "right" in the figure. To avoid conflicts, then, we need to use tristate gates in conjunction with a line that establishes the direction in which the data goes: 1 for left to right and 0 for right to left. Suppose that, in addition, we want to implement an enable line so that when it is 1, it allows the transceiver to work normally, but when it is 0, it disables *any* data transfer. Use any gates you deem necessary to implement this communication protocol.

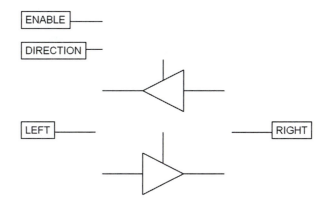

Figure 2.7. Circuit for Problem 14.

15. In serial communications the data bits are preceded by a "start bit," which is a high-voltage (5 V) bit. The data then follows the start bit. After the data comes the parity bit. The parity bit is set according to the following protocol: If it is even (odd), then the total number of 1's in the data *plus* the parity bit is even (odd). Suppose that we decide to send the sequence 1011001. If a logic 1 is encoded with 5 V (H) and a logic 0 is encoded by 0 V (L), then the data stream will be HHLHHLLHP, where P stands for the parity bit and the first bit is the start bit.

    (a) If the parity protocol is odd, what logic level (H/L) should the parity have? Justify your answer.

    (b) Baud rate is the number of bits per second sent in serial communications. If the baud rate is 19,200, what it the temporal width of each bit being transmitted?

    (c) Draw the waveform for sending the data stream 0110110 with even parity. Label your axes.

    (d) The USB communications protocol is different from the usual serial communications protocol. It uses the *Non Return to Zero Invert* (NRZI) encoding. In this protocol, the logic information is encoded via *changes* of voltage levels. A logic 0 corresponds to a change in the input voltage while a logic 1 corresponds to no change in the voltage level. For example, 0110 is encoded by Start(H)-L-L-L-H, where we have used H for high voltage and L for low voltage. In the NRZI encoding, there is no parity bit. Instead, the one's complement of the data is sent immediately after the data. Redraw the waveform of the previous part (c) using NRZI encoding.

16. In the following table, fill in the logic level interpreted by the gate of a given type for the voltages listed in the column on the left.

Input Voltage (V)	TTL-LS	CMOS-HC	CMOS-HCT
3.6			
0.2			
4.5			
−0.1			
1.4			
2.6			

17. Fill in the table with a "yes" or "no" depending on whether the gates of the type listed can output the voltage given in the column on the left.

Output Voltage (V)	TTL-LS	CMOS-HC	CMOS-HCT
0.1			
3.6			
2.1			
5.0			
0			
0.5			

2.8   LAB PROJECTS

This section presents ideas for lab activities.

2.8.1   Serial Transmission of ASCII-Coded Characters

Required equipment: PC with a serial port (see Figure 2.8), serial port connector with pins 2 and 3 tied up, oscilloscope and probe, and serial-port driver program.

Description   Serial transmission of digital data is an important component of computer communications. The Ethernet system uses this mode of transmission of data. It consists of sending the information bit by bit in a sequence. In this lab, you see how this is done for a popular communications protocol called the RS-232. This type of communication is used when a PC communicates to peripherals. It is becoming increasingly obsolete today and is being replaced by another new serial-port technology: Universal Serial Bus (USB). The difference in the new system is that the electronics at both ends are smarter. In turn, the series communications packets are complex. For our purposes, the more basic RS-232 is better because it transmits bytes of data in a simple way to recognize.

The RS-232 "protocol" works the following way: It uses two electrical lines to send the data back and forth and a series of other lines for control. One device tells the other that is it ready to send data, and the receiver responds with a signal indicating that it is ready to receive the data. This is called "handshaking."

Most PCs still have a serial port that supports the RS-232 protocol. In the PCs, this port is a nine-pin "female" connector. Pins 7 and 8 are ground. Pins 2 and 3 are to transmit and receive data, respectively. Some of the other pins are for the handshaking. Table 2.21

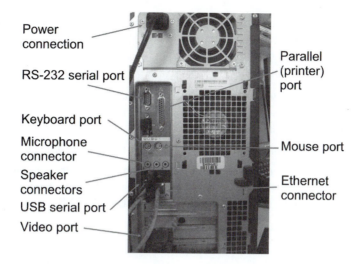

Power connection

RS-232 serial port

Keyboard port

Microphone connector

Speaker connectors

USB serial port

Video port

Parallel (printer) port

Mouse port

Ethernet connector

Figure 2.8. Back side of a typical computer from 2006. It shows the various I/O ports.

shows the pin definitions. Here we see how data is sent serially by "faking" the PC. That is, connecting pins 2 and 3 ensures that the PC receives what it sends, and connecting the other control lines (7–8 and 4–6) ensures that the PC registers that all is fine in the communication.

Table 2.21. Pin Connections of RS-232 Protocol for the Serial Port

Description	Name	Pin	I/O
Transmit data	TD	3	Output
Receive data	RD	2	Input
Request to send	RTS	7	Output
Clear to send	CTS	8	Input
Data terminal ready	DTR	4	Output
Data set ready	DSR	6	Input
Data carrier detect	DCD	1	Input
Ring indicator	RI	9	Input
Signal ground	SG	5	

Procedure   Plug the nine-pin connector with the rigged transmission lines to the serial port of your PC (port A). Connect pin 2 of the connector to channel 1 of the scope, and connect pin 7 to ground. Enter the communications settings into the serial driver program: COM1 port, 9,600 baud rate, even parity, and 7 bits. Measure the waveform that results after pressing a key.

Serial transmission follows a certain protocol:

- The data is preceded by a high-voltage start bit.

- Seven bits follow with the ASCII code.

- The last bit is a "parity" bit, used as a check for any errors in the transmission. When the parity is set to even, the parity bit is set accordingly so that the total number of 1s in the character *and* the parity bit is an even number. That is, it will be set to 1 when there is an odd number of 1s in the character.

Questions:

1. Determine how the data is transmitted. This entails figuring out the logic levels (0 = high voltage, 1 = low-voltage, or vice versa), and the order in which the bits are transmitted (MSB first or last).

2. Measure the bit transmission frequency and identify the units of the baud rate.

3. Show an example of the protocol by picking a character to transmit and making a detailed drawing of what you will observe on the oscilloscope when you press that key. Identify also the meaning of each bit (start bit, MSB, LSB, parity bit).

2.8.2 Practicum Test

You are given an IC that contains 3-input gates. We have painted over the label so that you do not know the type of gate it is.

1. The unknown IC has the standard pin connections for power and ground. Wire them.

2. Pins 1, 2, and 13 are inputs, and pin 12 is an output. Determine the function of the gate. Explain your procedure in making your determination.

CHAPTER 3

COMBINATIONAL LOGIC

Contents

Combinational logic refers to the type of digital operations in which the information is processed in parallel. That is, the output of the circuit depends on the present values of the inputs. This is in contrast to *sequential logic*, in which the output of the circuit may depend on the value of variables at a previous time. We start by formally introducing *Boolean algebra* and then move to the simplification of functions, which is central to combinational logic.

3.1 BOOLEAN ALGEBRA

This is an algebraic structure defined on a set of elements $B = \{0, 1\}$ together with two binary operators, $+$ (OR) and $\cdot$ (AND). The postulates of this structure are:

1. **Closure.** B is closed with respect to OR and AND. Thus, given $x, y \in B$, then $x + y = z \in B$ and $x \cdot y = w \in B$.

2. **Identity.** Identity elements exist, one with respect to OR, and another one with respect to AND. For the OR operation, the identity is 0:

$$x + 0 = x \tag{3.1}$$

   For the AND operation, the identity is 1:

$$x \cdot 1 = x \tag{3.2}$$

3. **Commutative.** The OR and AND operators are commutative. This means that the following relations hold:
$$x + y = y + x \tag{3.3}$$

   and
$$x \cdot y = y \cdot x \tag{3.4}$$

4. **Distributive.** The OR operation is distributive over AND:

$$x \cdot (y + z) = x \cdot y + x \cdot z \tag{3.5}$$

   The AND operation is distributive over OR:

$$x + (y \cdot z) = (x + y) \cdot (x + z) \tag{3.6}$$

   The latter is an important difference between Boolean algebra and the algebra of real numbers $[3 + 1 \cdot 2 \neq (3 + 1) \cdot (3 + 2)]$.

5. **Complement or inverse.** The OR and AND operators have an inverse: If $x \in B$ then $\exists\, x' \in B$ such that
$$x + x' = 1 \tag{3.7}$$

   and
$$x \cdot x' = 0 \tag{3.8}$$

3.2 THEOREMS

The theorems listed here (without proof) follow the postulates of Boolean algebra:

$$(A')' = A \tag{3.9}$$
$$A \cdot A = A \tag{3.10}$$
$$A' + A = 1 \tag{3.11}$$
$$A' \cdot A = 0 \tag{3.12}$$
$$A + 0 = A \tag{3.13}$$

$$A + 1 = 1 \tag{3.14}$$

$$A \cdot 0 = 0 \tag{3.15}$$

$$A \cdot 1 = A \tag{3.16}$$

An important theorem in Boolean algebra is *De Morgan's theorem*. The theorem can be expressed as follows:

$$(A + B)' = A' \cdot B' \tag{3.17}$$

or

$$(A \cdot B)' = A' + B' \tag{3.18}$$

NOTATION ALERT: For denoting the AND operation, we use the same notation shorthand as multiplication of real numbers: $xy = x$ AND y.

3.3   NAND-GATE IMPLEMENTATION

We can use De Morgan's theorem to convert a function or circuit composed of OR and AND gates to one with just NAND gates. This conversion makes the electronic implementation of the circuit economical and efficient.

Consider this Boolean function:

$$F = AB' + A'B$$

Such a function is implemented by a digital circuit that uses two NOT gates, two AND gates, and one OR gate, as shown in Figure 3.1.

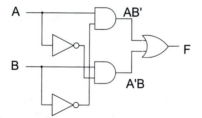

Figure 3.1.   Circuit diagram for implementing function F (see text).

If we take the complement of the function F twice:

$$F = (F')'$$

which becomes:

$$F = ((AB' + A'B)')'$$

Applying De Morgan's theorem to the inner parenthesis, we get

$$F = ((AB')'(A'B)')'$$

which represents three nested NAND gates. The NOT gates can be replaced by NAND gates, too, so the circuit will look as shown in Figure 3.2. This implementation makes the wiring of the circuits more straightforward and economical.

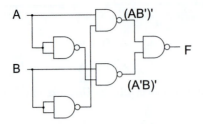

Figure 3.2.   Implementation of function F (see text) using only NAND gates.

3.4   REPRESENTATION OF BOOLEAN FUNCTIONS

Here we discuss equivalent ways to represent Boolean functions. These can be analytical, tabular, or graphical.

3.4.1   Analytical

We can represent functions in two different algebraic forms: As *sums of products* (SOP), or as *products of sums* (POS). The most common type is the former, owing to its resemblance to the algebraic structure of real numbers. For example, a function expressed in an SOP form looks like this:

$$F = x + yz + x'z$$

A function expressed as a POS looks like this:

$$F = (x + y')(x' + y' + z)(y + z')$$

3.4.2   Tabular

A direct way to represent a function is to use a table listing the values of the function for each combination of the input variables. For example, a function F of three variables, x, y, and z can be defined by the truth table shown in Table 3.1. Note that there are 8, or 2^3, possible combinations of 0's and 1's. If we have n variables, there are 2^n distinct combinations of values. An SOP function is implemented when the input variables have values for which the function is 1. For example, in the case of Table 3.1, $F = 1$ when $x = 0$, $y = 0$, and $z = 1$. The product $x'y'z$ is 1 when the variables take those values. If we find the products that give 1 for the other combinations in the table and OR them together, we would have converted truth-table information into the SOP form. In the next section we discuss how to do this in a systematic way.

   To implement the POS form of the function using Table 3.1 we focus on the combinations for which $F = 0$, such as $x = 0$, $y = 1$, and $z = 1$. The term $x' + y + z$ will be 0 for that combination of values. If we find the other "sums" for which $F = 0$ and AND them together, we would have converted the tabular information into a POS analytical form.

■ **EXERCISE 3.1**

   Use truth tables to verify the theorems of Equations 3.9–3.16.

Table 3.1.   Example of a Truth Table for a Function of Three Variables

x	y	z	F
0	0	0	0
0	0	1	1
0	1	0	1
0	1	1	0
1	0	0	1
1	0	1	0
1	1	0	1
1	1	1	1

■ **EXERCISE 3.2**

Convert the information in Table 3.1 into an SOP function.

3.4.3 Graphical

We can also represent a function in a graphical way as a Venn diagram. For example, the function $F = xz + yz$ can be expressed by the shaded areas in the diagram of Figure 3.3.

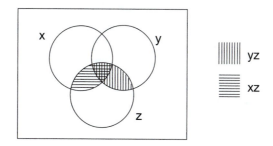

Figure 3.3.   Venn diagram for three Boolean variables. Shaded areas represent function F in the text.

3.5 SIMPLIFICATION OF FUNCTIONS

Whenever we want to implement a Boolean function in a digital circuit, we want the circuit to be the simplest possible. The overriding factor for this is economic. In particular, we want to have a minimum number of gates and a minimum number of inputs per gate. If a Boolean function is not in its simplest form, we may want to minimize it. Next, we discuss three methods for simplifying functions.

3.5.1 Algebraic

Here we apply the postulates and theorems of Boolean algebra to simplify a given function. For example, consider this function:

$$F = x'y'z + x'yz + xy'$$

We can simplify it by factorizing $x'z$ in the first two terms

$$F = \underbrace{x'zy' + x'zy} + xy'$$

or

$$F = x'z(y' + y) + xy'$$

By the use of Theorem 3.11 we get

$$F = x'z + xy'.$$

However, the previous simplification is an easy one. Other expressions are not as straightforward to simplify because they require the use of the distributive law of OR over AND. We are not used to doing this type of simplification in ordinary algebra. Consider, for example, the function:

$$F = x' + xy$$

If we use the distributive law, we get

$$F = (x' + x)(x' + y)$$

which simplifies to

$$F = x' + y$$

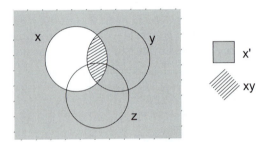

Figure 3.4. Using the Venn diagram to simplify functions.

3.5.2 Graphical

We simplify the previous example visually by using the Venn diagram of Figure 3.4. After shading the individual terms, we can re-express the shaded areas with a simpler expression:

$$F = x' + y$$

Because this method is visual, it is easier than the algebraic one. Its drawback is the time required to draw the Venn diagram and shade the corresponding regions.

■ **EXERCISE 3.3**

  Simplify the following function: $F = xyz + x'y + xyz'$.

■ **EXERCISE 3.4**

  For function $F = zx + zx'y$,

  1. *Simplify it algebraically (if out of ideas try a Venn diagram).*
  2. *Draw a circuit diagram using AND and OR gates.*
  3. *Draw a circuit diagram using only NAND gates.*

The best way to simplify functions is to use a more efficient variation of the Venn diagram method, known as the Karnaugh map method. We discuss this method in great detail in the next section.

3.6 KARNAUGH MAPS

The most important aspect of combinational logic is the simplification of Boolean functions. The algebraic method can be cumbersome and inefficient. In contrast, the Karnaugh map method provides a simple and fast way to simplify functions.

3.6.1 Minterms

As mentioned earlier, numbers with n binary digits have 2^n unique combinations of 1's and 0's. The minterm is a Boolean function that uniquely determines each combination. It is defined as a one-term function that is 1 when all the variables or their complements are ANDed together. Table 3.2 gives the minterms for each of the combinations of 3-bit binary numbers.

Table 3.2. Minterms of Three Variables

x	y	z	Minterm	Designation
0	0	0	$x'y'z'$	m_0
0	0	1	$x'y'z$	m_1
0	1	0	$x'yz'$	m_2
0	1	1	$x'yz$	m_3
1	0	0	$xy'z'$	m_4
1	0	1	$xy'z$	m_5
1	1	0	xyz'	m_6
1	1	1	xyz	m_7

For example, consider the minterm of the first row: $z'y'z'$. This function is a 1 when $x = 0$, $y = 0$, and $z = 0$. It is 0 for all other combinations. The same is the case for the other minterms: They are 1 only for the corresponding combination of variables. Minterms are useful because they set up the simplification of a function when the starting point is the truth table. For example, consider the truth table in Table 3.3. We can represent the function F by ORing all the minterms for which the function is a 1:

$$F = x'y'z' + x'yz' + xy'z' + xyz' + xyz \tag{3.19}$$

This way, each minterm is 1 only when the input variables have the corresponding combination of values. A truth table can be constructed easily when a function is expressed as a sum of minterms.

■ **EXERCISE 3.5**

Write the truth table for the function $F = x'yz + xy'z + xyz$.

Table 3.3.   Example of a Truth Table for Three Variables

x	y	z	F
0	0	0	1
0	0	1	0
0	1	0	1
0	1	1	0
1	0	0	1
1	0	1	0
1	1	0	1
1	1	1	1

3.6.2 Two-Variable Map

How do we best simplify a Boolean expression such as the one of Equation 3.19? As it stands now, the function requires five 3-input AND gates and one 5-input OR gate. We can simplify it algebraically, as done in the previous section, but instead here we use the Karnaugh map method, which is more efficient. We start answering this question by considering the case of two variables. If we have two variables x and y, there are four possible combinations of binary values between them, and therefore, four minterms: $x'y'$, $x'y$, xy', and xy. If we draw the corresponding Venn diagram, it also has four areas, with each continuous region corresponding to a minterm, as shown in Figure 3.5. The Karnaugh map (K-map) is a more structured form of the Venn diagram. The two-variable K-map is shown in Figure 3.6.

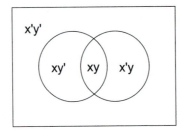

Figure 3.5.   Venn diagram for two variables with minterms assigned to each of the four sectors.

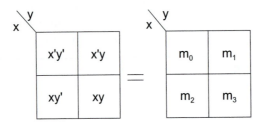

Figure 3.6.   Karnaugh map for two variables.

The structure of this map makes it easy to identify the various combinations of regions. For example, you can quickly see that the lower two boxes correspond to the variable x, as shown in Figure 3.7. We can show this algebraically by ORing the minterms that correspond to each box:

$$xy' + xy = x(y' + y) = x$$

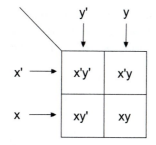

Figure 3.7. Identifying the domain regions of each variable in the two-variable Karnaugh map.

Similarly, you can see that the top two boxes are x', the right two boxes are y, and the left two boxes are y'. We can then conclude that any two *contiguous* boxes grouped together form a Boolean term of one variable.

Let us move now to simplify functions. Consider the function $F = x'y' + xy'$. Although it is straightforward to solve this analytically, let us do it using the K-map method. The procedure consists of the following steps:

1. Put a 1 in the minterm boxes that make up the function, leaving the others blank. If the starting point is a truth table, put a 1 in the boxes that correspond to combinations where the function is a 1.

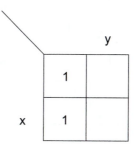

2. Group any contiguous boxes with a closed loop, as shown, and identify the Boolean term that the loops represent. This is equivalent to hatching the areas in the Venn diagram.

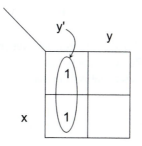

Our answer is therefore $F = y'$.

In the next situation we start from the truth table. The map is arranged like a grid, with rows and columns representing the values of the variables. Therefore, to find a particular combination of variables, we look for the box with coordinates that match.

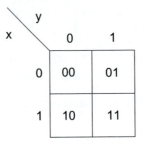

As an example, consider the following truth table:

x	y	F
0	0	1
0	1	1
1	0	1
1	1	0

We then draw a map with the rows and columns labeled on the outside of the map, as shown in the previous map, and put a 1 in the boxes that correspond to the combinations 00, 01, and 10. Then we group any contiguous boxes, as shown next.

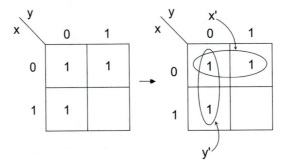

3. Finally, we write the simplified function as the algebraic terms that represent each group of looped boxes ORed together:

$$F = x' + y'$$

Notice that in simplifying the function we have made it a bit redundant: the term $x'y'$ is part of both loops. Redundancy is fine in this case since the function has a simpler form and will require fewer gates to implement.

■ **EXERCISE 3.6**

Simplify the function $F = xy + x'y + x'y'$.

3.6.3 Three-Variable Map

Because three variables can form eight possible combinations of 1's and 0's, the corresponding K-map must have eight boxes. A one-variable term now occupies four contiguous boxes, while two-variable terms occupy two contiguous boxes. Figure 3.8 shows one representation of the three-variable map. Notice that we have disrupted the order of the binary numbers that label the columns: We do 00, 01, 11, 10 instead of 00, 01, 10, 11. This is done so that variable y occupies four contiguous boxes, making it easier to reduce functions.

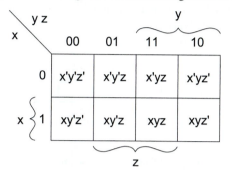

Figure 3.8. Karnaugh map for three variables.

For example, suppose that a function consists of four contiguous boxes with coordinates (or combinations) 000, 001, 100, and 101. The function can be expressed as

$$F = x'y'z' + x'y'z + xy'z' + xy'z$$

Analytically, it can be simplified to

$$
\begin{aligned}
F &= x'y'(z' + z) + xy'(z' + z) \\
&= x'y' + xy \\
&= y'(x' + x) \\
&= y'
\end{aligned}
$$

which is in agreement with the labeling of the preceding K-map.

The three-variable map has been designed so that the main three variables, x, y, and z each occupy four contiguous boxes. The only exception is z', which is defined by wrapping around the left-most and right-most columns of the K-map.

As an example, consider the function $F = x'y'z + x'yz + xyz + xyz'$. Using the K-map, it can be simplified to:

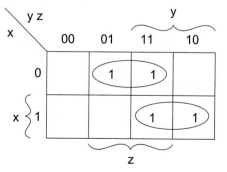

The result is $F = x'z + xy$.

■ **EXERCISE 3.7**

Show that the function given by Equation 3.19 reduces to $z' + xy$. By simplifying the function, we now need only three gates to implement it.

3.6.4 Four-Variable Map

The four-variable map follows the same prescriptions as the other maps, with now, one-variable term occupying eight boxes, and two- and three-variable terms occupying four and two contiguous boxes, respectively. Figure 3.9 shows the full map. Notice that now the rows are ordered in the same way as the columns.

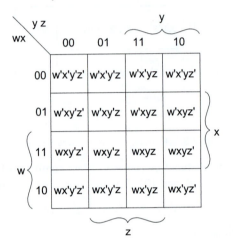

Figure 3.9. Karnaugh map for four variables.

As an example, let us simplify this function:

$$F = wxy + yz + xy'z + x'y$$

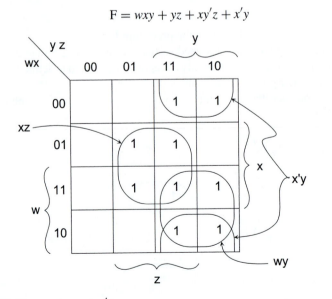

The result is $F = xz + wy + x'y$.

■ **EXERCISE 3.8**

Simplify $F = w'x'y'z' + wx' + x'y + xyz$.

3.6.5 Don't Care Conditions

Don't care conditions are combinations of input variables that will *never* occur. This situation arises often in combinational logic. We can then use this information to our advantage by using the boxes that correspond to those combinations as wildcards. That is, we may or may not include them when grouping boxes as part of the simplification. Consider the following problem. WXYZ and ABCD are binary-coded decimals (BCD) in this operation:

$$
\begin{array}{cccc}
\text{W} & \text{X} & \text{Y} & \text{Z} \\
+ & & & 1 \\
\hline
\text{G} \quad \text{A} & \text{B} & \text{C} & \text{D}
\end{array}
$$

(Remember that BCD digits are never higher than 9.) We want to construct a circuit that simulates adding 1 to a decimal digit. As shown earlier, the input number has digits W, X, Y, and Z. The output number has digits A, B, C, D, and the carry G. We implement this by making the output digits and the carry be functions of the input digits. We define the truth table for each of them by doing all the operations that may occur beforehand (by brute force), as shown:

W	X	Y	Z	G	A	B	C	D
0	0	0	0	0	0	0	0	1
0	0	0	1	0	0	0	1	0
0	0	1	0	0	0	0	1	1
0	0	1	1	0	0	1	0	0
0	1	0	0	0	0	1	0	1
0	1	0	1	0	0	1	1	0
0	1	1	0	0	0	1	1	1
0	1	1	1	0	1	0	0	0
1	0	0	0	0	1	0	0	1
1	0	0	1	1	0	0	0	0
1	0	1	0	×	×	×	×	×
1	0	1	1	×	×	×	×	×
1	1	0	0	×	×	×	×	×
1	1	0	1	×	×	×	×	×
1	1	1	0	×	×	×	×	×
1	1	1	1	×	×	×	×	×

Here the "×" denotes a "don't care" condition. It represents a combination of input variables that will not occur, so we don't care whether it is a 1 or a 0.

As an example, we solve for A and B:

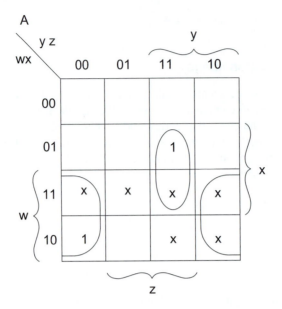

The result is $A = wz' + xyz$. Notice that we group some of the don't care symbols with 1s so that we end up with a simpler Boolean term. For B, we have this:

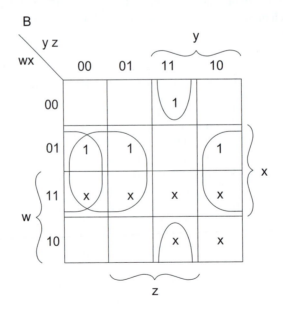

The result is $B = xy' + xz' + x'yz$.

■ EXERCISE 3.9

A function F is true when the 4-bit binary-coded decimal input is a 3 or a 5. Find the simplest expression for F.

3.7   MORE THAN FOUR VARIABLES

It is reasonable to do the previous simplifications by hand for up to four variables. For more variables it is no longer practical to continue with the same method.

3.7.1   Three-Dimensional Karnaugh Maps

We can keep going with Karnaugh maps for two more variable increments: five and six. However, our Karnaugh map becomes three-dimensional, with the six-variable map resembling a Rubik's cube, but with four boxes on the side. It is obvious that the practicality of the method goes down for more than four variables. Beyond four variables, we can use the Quine-McCluskey method of prime implicants, which is more suitable for solving with a computer. However, it is worth noting that increasing the variables increases the computation time exponentially. We will not cover these methods here.

3.7.2   Brute–Force Logic

If minimizing becomes troublesome, we can go back to brute–force. It does not sound very appealing. However, technology has made this a worthwhile option. The programmable logic array (PLA), complex programmable logic device (CPLD), and field programmable logic array (FPGA) are advanced circuits that have arrays of gates all within a chip that can be programmed. In the next chapter we will see a version of this in programmable read-only memory (PROM). In fact, if the final product is a sophisticated circuit, or if we are making identical copies of the same complex circuit, we would consider these advanced options. Because these options lump everything within a chip, they do not help us learn electronics, so we do not present an in-depth discussion of these devices.

3.8   WRAP-UP

In summary, when we have a digital function of four variables or less, we can successfully obtain a simplified expression of the function by using the Karnaugh map method. The goals of this simplification method are listed here:

1. Minimize the number of gates.
2. Minimize the number of inputs per gate.

The procedure for solving a given problem involves the steps listed below:

1. State the problem.
2. Set the input and output variables.
3. Complete the truth table.
4. Simplify the function using the Karnaugh map method.
5. Convert the circuit to one that uses only NAND gates.
6. Draw the corresponding circuit diagram.

3.9   WIRING DIGEST: OPEN COLLECTOR/DRAIN OUTPUTS

The technology involved in making the gates allows a type of output that is convenient in certain applications. This type of output is called *open collector* for TTL-based circuits and *open drain* for CMOS-based circuits. For example, Table 3.4 gives the number of logic ICs that are open collector/drain. Electronically, the output of a gate is connected to either the collector of a BJT transistor or the drain of a MOSFET transistor. Although we discuss these components in detail in Chapter 11 (Connecting Digital to Analog and to the World), suffice it to say that these devices are used as electrical switches. The collector or drain to which the output is connected gets switched to ground when the logic output is 0, as shown in Figure 3.10. When the output is a logic 1, then that transistor switch gets turned off, but additional circuitry takes care of putting the right voltage level on the line. One use is to have the output line connected only to the collector/drain of the gate. In this case, when the output is a logic 1, the transistor switch is open and the output is effectively disconnected.

Table 3.4. Open Collector/Drain ICs

IC Type* #	Description
01, 03†, 26#	2–input NAND
05, 06‡, 16#	NOT
07, 17#	BUFFER
09	2–input AND
12	3–input NAND
15	3–input AND
22	4–input NAND
47#	BCD to 7 segment decoder

*Open-drain numbers are preceded by "1G." †Different pin-out than 01.
#Up to 15-V TTL outputs. ‡Up to 30-V TTL outputs.

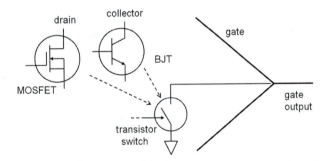

Figure 3.10. Qualitative schematic of the internal wiring of an open collector/drain output.

To wire open-collector gates, we must supply a high voltage when we have a logic 1. In such a case, we use a "pull-up" resistor, as shown in Figure 3.11. A 1 kΩ resistor suffices. Why do we want to use open-collector outputs in the first place? In early days of digital circuits, they were used as an easy way to connect gate outputs to a single line. When we wanted to disable a particular output, it was set to a logic 1. Two open-collector gates wired together produce the same effect as if the lines were fed into an AND gate, a configuration that is also known as *wired-AND*. That is, the voltage level of the line is high only if all the outputs are logic 1's; otherwise, the voltage level is low. This AND function is for positive

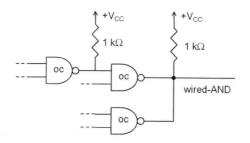

Figure 3.11.   Example showing how open-collector/drain gates are connected.

logic. In negative logic, it works like an OR, so it is called *wired-OR*. These applications are not common today because of their low speed; tristate buffers and drivers are a faster replacement. Currently, this configuration has a new and different use: To drive gates that work at different voltage levels, such as those in the LV technology.

3.10 PROBLEMS

1. Consider the following functions:

    (a) $F = x'y' + xy + x'y$
    (b) $F = (x + y)(x + y')$
    (c) $F = x' + xy + xz' + xy'z'$

    - Reduce them to the simplest expression using the postulates of Boolean algebra.
    - Check your results using a Venn diagram.
    - Draw a circuit diagram of the simplest expression of the function.

2. Find a simplified expression for $F = a \odot (a \odot b)$.

3. Verify De Morgan's theorem using a truth table: Make a table with two columns listing all the possible combinations of the two inputs, A and B. Fill additional columns with the result of the operations of each side of Equations 3.17 and 3.18 for each combination of A and B.

4. In negative logic, a logic 1 is a low voltage and a logic 0 is a high voltage. Use De Morgan's theorem to show that, in negative logic, an AND gate based on positive logic behaves as an OR gate, and vice versa.

5. A CNOT gate has two inputs, a and b, and two outputs, a and $a \oplus b$. The right part of Figure 3.12 shows the symbolic representation, and the one on the left side shows the wiring diagram. Briefly, one of the outputs is the same as one of the inputs (a). The other output is the XOR operation between the two inputs ($c = a \oplus b$). This gate is the inverse of itself.

    (a) Use a truth table to show this property by proving that $(\text{CNOT})^2 = I$, where I is the identity operation (the outputs are the same as the inputs).
    (b) The CNOT gate has two inputs and two outputs. Find the 4×4 matrix that represents it.
    (c) Show by matrix multiplication that $(\text{CNOT})^2 = I$.

Figure 3.12.   Diagram of the CNOT gate (Problem 5).

6. A Toffoli gate is said to be a universal gate. It has three inputs and three outputs. Figure 3.13 shows its circuit diagram. If the three inputs are a, b, and c, the three outputs are a, b, and $t = (ab) \oplus c$. Design an arrangement of input/output connections so that the Toffoli gate can be used as follows:

(a) An AND gate (two inputs, x and y, and an output, xy)

(b) A NAND gate

(c) An OR gate

Suggestion: Review the definition of XOR. You do not need to use all three outputs of the Toffoli gate. For example, to construct a NOT gate with a Toffoli gate, we set input $c = 1$ and tie inputs a and b. The input to the NOT gate is $i = ab$, and the output is $t = 1 \oplus i$.

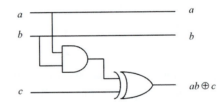

Figure 3.13.   Diagram of the Toffoli gate (Problem 6).

7. A 1-bit memory cell consists of a latch and a tristate buffer, as shown in Figure 3.14. It uses a line for both input and output. When $\overline{SEL} = 0$, the latch captures the logic level of the line at IN and transfers it to OUT. When $\overline{SEL} = 1$, the logic value of OUT remains the same regardless of the value on IN. The value at OUT can be connected to the INPUT/OUTPUT line by making $T = 1$ in the tristate buffer, enabling it. When $T = 0$, the tristate buffer is disabled and isolates OUT from the INPUT/OUTPUT line. We control the memory cell via two control lines: $\overline{ENABLE}$ and $READ/\overline{WRITE}$. The communications protocol is then the following:

- When we wish to communicate with the memory cell we make $\overline{ENABLE} = 0$. In addition,
  - When we want to write to the memory cell, we make $READ/\overline{WRITE} = 0$. When this occurs, we need to enable the latch and disable the buffer.
  - When we want to read the value stored in the memory cell, we make $READ/\overline{WRITE} = 1$. Then we need to disable the latch and enable the buffer.
- When we do not want to communicate with the memory cell, we need to disable the buffer but still make sure that the memory remembers its stored value (keeping OUT from changing).

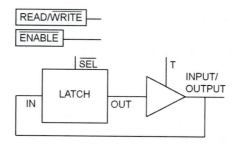

Figure 3.14. Diagram of the circuit for driving a 1-bit memory cell (Problem 7).

(a) Based on the specified protocol, fill in a truth that has $\overline{ENABLE}$ and $READ/\overline{WRITE}$ as inputs and $\overline{SEL}$ and T as outputs.

(b) Find simplified expressions for $\overline{SEL}$ and T using Karnaugh maps.

(c) Implement the memory cell by completing the circuit of Figure 3.14. Use any gates you deem appropriate.

8. A three-member congressional board, made of members A, B, and C, is to vote on resolutions. A and B are with the ruling party, and C is with the opposition. C shares some interests with A and B, but not both. The board decided to save taxpayers money by using a voting machine that makes legislating more efficient.

(a) Write a truth table for the function F implemented by the voting machine, which has the following rules:

- When A and B vote yes and C votes no, the resolution passes.
- A, B, and C never all agree on anything (they never all vote yes). (Of course!)
- On certain environmental issues, A and C agree, but B disagrees. If A and C vote yes and B votes no, the resolution passes.
- On abortion rights, B and C agree, but A disagrees. If B and C vote yes and A votes no, then the resolution passes.
- The inputs are low active, so a yes is a logic 0 and a no is a logic 1.
- The output is high active: If the resolution passes, the output is a logic 1.

(b) Simplify F using Karnaugh maps.

(c) Draw a simplified circuit using only NAND gates.

9. Simplify the circuit of Figure 3.15 and redraw it using AND, NOT, and OR gates.

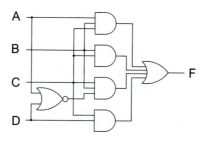

Figure 3.15. Circuit for Problem 9.

10. Table 3.5 shows the truth table for a "4-to-2 priority encoder."

Table 3.5. Truth Table for the Priority Encoder of Problem 10

Inputs[#]				Outputs	
D_3	D_2	D_1	D_0	S_1	S_0
0	0	0	×	0	0
0	0	1	×	0	1
0	1	×	×	1	0
1	×	×	×	1	1

[#]The symbol × can be 0 or 1.

(a) Write the complete truth table for S_1 and S_0 (all 16 combinations of D_3, D_2, D_1, and D_0).

(b) Draw the corresponding Karnaugh map.

(c) Obtain simplified Boolean expressions for them.

11. We want to generate the circuit of the seven-segment display decoder, the 7447. It has four inputs representing a BCD number. The outputs to the LEDs are *low active*. That is, they are 0 if we want to turn on the segment. Eventually, we want the seven segments to display the decimal digits from 0 to 9. For each segment:

(a) Write the truth table. Remember that a 0 turns on the segment light. For example, if F = 0101, we want to display a 5, which means that $\bar{a} = \bar{c} = \bar{d} = \bar{f} = \bar{g} = 0$ and $\bar{b} = \bar{e} = 1$.

(b) Draw the K-maps and obtain the simplest expression.

(c) Draw the circuit diagram using NAND gates.

12. We want to design a BCD-to-XS3 binary code converter. A number in XS3 (excess 3) code, N_{XS3}, is related to a BCD number, N_{BCD}, by:

$$N_{XS3} = N_{BCD} \text{ plus } 3$$

For example, the BCD number 1000 is 1011 in XS3.

The converter we want to design will have the digits of the BCD number (B_3, B_2, B_1, and B_0) as inputs and the digits of the XS3 number (X_3, X_2, X_1 and X_0) as outputs. It will work in such a way that when we input a BCD number at the B_i's, we get its XS3 equivalent at the outputs X_i's.

(a) Write the truth table for X_3, X_2, X_1, and X_0 as a function of B_3, B_2, B_1, and B_0.

(b) Find a simplified Boolean expression for X_2 and X_1.

(c) Draw a circuit diagram for X_2 and X_1 in terms of NAND gates.

13. Design a combinational circuit that has as inputs the 4-bit (binary) month count from a "month" counter, and the following as outputs:

(a) The 8 bits of the BCD representing the month (for example, for December the input is 1100_2, and the output is 00010010_2).

(b) Two functions, F_{30} and F_{31}, whose output is a 1 when the month count corresponds to a month with 30 and 31 days, respectively.

The design should include the following:

(a) Truth tables reflecting the statement of the problem.

(b) Karnaugh maps for each of the outputs.

(c) Circuit diagram using NAND gates.

14. One of the computer's most important parts is the device that performs all the arithmetic and logical operations: the arithmetic logical unit (ALU). The ALU is a large combinational circuit that has control inputs, data inputs, and data outputs. In this exercise, we use the 74LS181 4-bit ALU. The functional details are described shortly. The data inputs are $B = B_3B_2B_1B_0$ and $A = A_3A_2A_1A_0$, and the data outputs are $F = F_3F_2F_1F_0$. They represent 4-bit binary variables, with F being the result of an operation between A and B. The control inputs S_3, S_2, S_1, S_0, M, and C_n specify the operation as given in Table 3.6. Note that the functions under "$M = 1$" are independent of the input C_n.

Table 3.6. Functional Table of the ALU

S_3 S_2 S_1 S_0	$M = 1$	$M = 0, C_n = 1$	$M = 0, C_n = 0$
0 0 0 0	$F = A'$	$F = A$	$F = A$ PLUS 1
0 0 0 1	$F = (A + B)'$	$F = A + B$	$F = (A + B)$ PLUS 1
0 0 1 0	$F = A'B$	$F = A + B'$	$F = (A + B')$ PLUS 1
0 0 1 1	$F = 0$	$F = $ MINUS 1	$F = $ ZERO
0 1 0 0	$F = (AB)'$	$F = A$ PLUS AB'	$F = A$ PLUS AB' PLUS 1
0 1 0 1	$F = B'$	$F = (A + B)$ PLUS AB'	$F = (A + B)$ PLUS AB' PLUS 1
0 1 1 0	$F = A \oplus B$	$F = A$ MINUS B MINUS 1	$F = A$ MINUS B
0 1 1 1	$F = AB'$	$F = AB'$ MINUS 1	$F = AB'$
1 0 0 0	$F = A' + B$	$F = A$ PLUS AB	$F = A$ PLUS AB PLUS 1
1 0 0 1	$F = A \odot B$	$F = A$ PLUS B	$F = A$ PLUS B PLUS 1
1 0 1 0	$F = B$	$F = (A + B')$ PLUS AB	$F = (A + B')$ PLUS AB PLUS 1
1 0 1 1	$F = AB$	$F = AB$ MINUS 1	$F = AB$
1 1 0 0	$F = 1$	$F = A$ PLUS $A^\dagger$	$F = A$ PLUS A PLUS 1
1 1 0 1	$F = A + B'$	$F = (A + B)$ PLUS A	$F = (A + B)$ PLUS A PLUS 1
1 1 1 0	$F = A + B$	$F = (A + B')$ PLUS A	$F = (A + B')$ PLUS A PLUS 1
1 1 1 1	$F = A$	$F = A$ MINUS 1	$F = A$

PLUS represents addition. MINUS represents subtraction.† Each bit is shifted to the left.

As you can see from Table 3.6, we can perform 48 operations with the 74LS181 ALU. However, suppose that for a given application we want to use only eight of those operations. Then having to provide input to the six control lines of the ALU is wasteful. Because there are eight possible combinations of 3 bits, the minimum number of variables that we need to specify the eight functions is three. In the following circuit, the control lines of the ALU are fed from a decoding circuit, which you should design. This circuit tells the ALU which operation to perform.

The decoding circuit has three input variables x, y, and z, that represent a "code." This code specifies which operation is to be performed. The role of the decoder circuit is to set the inputs of the ALU to the levels required for the corresponding operation, as shown in Figure 3.16. The first two sections of Table 3.7 give the correspondence between the input variables x, y, and z, and the operations that the ALU must perform. A completely filled third section would have the values that each ALU input requires for setting the proper ALU operation. We can think of the inputs of the ALU S_3, S_2, S_1, S_0, M, and C_n as Boolean functions of x, y, and z.

(a) Complete Table 3.7.

(b) Get a simplified functional form for each of the ALU control inputs as a function of x, y, and z, using the Karnaugh map method.

(c) Express each circuit in terms of NAND gates.

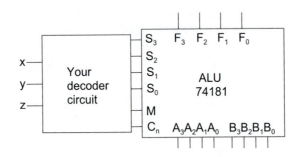

Figure 3.16.   Wiring of the circuit decoder to drive the ALU.

Table 3.7.   Truth Table for the Lines of the ALU as a Function of the Input Variables of the Decoder of Figure 3.16

x	y	z	ALU OPERATION	S_3	S_2	S_1	S_0	M	C_n
0	0	0	$F = A + B$	1	1	1	0	1	×
0	0	1	$F = A \cdot B$	1					
0	1	0	$F = A'$	0					
0	1	1	$F = B'$	0					
1	0	0	$F = A \oplus B$	0					
1	0	1	$F = A \odot B$	1					
1	1	0	$F = A$ minus B	0					
1	1	1	$F = A$ plus B	1					

3.11   LAB PROJECTS

3.11.1   The TTL Half Adder: Design and Construction

Lab credit: D. Glenar.

Required equipment: Logic board. (The logic board is a box that provides logic inputs to a circuit via switches, and displays the outputs of a circuit on light-emitting diodes, LED. For details on making one see Appendix A.)

Required components: 74LS00 (or 74HCT00) ICs, common-anode seven-segment display, resistor array chip, 74LS47 (or 74HCT4511) seven-segment driver.

NOTE: You must look at the pin connections and specifications of all chips at www.datasheetcatalog.com.

A First Circuit   The goal of this lab is to implement the half adder, shown schematically below. It performs the addition of two binary digits x and y yielding a sum bit S and a carry C.

The design of the half adder circuit involves filling in the truth table in Table 3.8 for the outputs S and C, finding a Boolean expression for S and C, and wiring a circuit that contains only NAND gates.

Does the circuit work? No? Figuring this out is an art. Appendix B gives you a few guidelines and hints.

Table 3.8.   Truth Table for the Half Adder

x	y	C	S
0	0		
0	1		
1	0		
1	1		

Displaying the Result with the Seven-Segment Display   Most digital displays use the seven-segment display. Figure 3.17 shows this type of display. The schematic of the display is shown in Figure 3.18. Each of the seven segments of the display is a different LED. The segments you will use are wired in the "common anode" configuration, as shown in Figure 3.18. To wire them, you need to put a 220 Ω resistor in series with each LED. However, because this is a common occurrence in digital, the seven needed resistors come in a single package. The package is set up so that the resistors connect the pins that are opposite to each other in the chip. Wire the resistor package to the seven-segment and test it.

Figure 3.17.   Photo of two seven-segment displays.

The BCD to Seven-Segment Decoder   For driving seven-segment displays we have two options: The TTL 74LS47, or its CMOS equivalents, 74HC4511 or 74HCT4511. These are decoders that have four inputs that represent a binary-coded decimal (BCD) number.

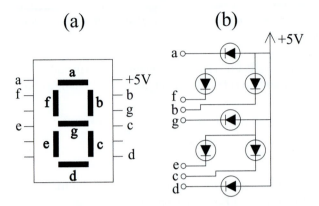

Figure 3.18.   Diagram of the wiring of a common-anode seven-segment display.

Table 3.9.   Steps to Drive the Seven-Segment Display Depending on the Digital IC Used

Step	TTL(74LS47)	CMOS (74CHT4511)
1	RBI (pin 4) and LT (pin 3) to +5 V.	RBI (pin 4) and LT (pin 3) to +5 V.
2	BI (pin 5) to +5 V (optional)	LE (pin 5) to a switch in the logic board.
3	The a, b, c, d, e, f, g outputs to the resistor unit.	The a, b, c, d, e, f, g outputs to the resistor unit.
4	Resistor unit to common-anode display.	Resistor unit to common-cathode display.
5	Data inputs (A3, A2, A1, A0) to switches in the logic board.	Data inputs (D3, D2, D1, D0) to switches in the logic board.

The pin connections for the decoder are given in the data sheets. The TTL and CMOS ICs are pin-compatible. The outputs labeled a, b, c, d, e, f, and g are functions whose output should light the LED segments. The decoders have a shortfall, though: They decode and display properly decimal numbers (from 0 to 9), but do not decode binary numbers above 9. The 74LS47 has low-active outputs: They output 0 to light the corresponding common-anode LED. Conversely, the CMOS 4511s are high active, so a common-cathode display must be used. A second difference that is important to us is that the CMOS decoder is latched, and updates the segment outputs only when the LE input goes from low to high. Table 3.9 gives the steps you must follow depending on which family of IC you use. Because our lab experiments are based on TTL ICs, we prefer to use the 74LS47 here. Wire the decoder and display according to the steps in the table. If 74LS47 are not available, then use the 74HCT4511 and wire it according to the table. Using the CMOS option has a consequence that is a nuisance: Every time you want to update the display you have to switch the LE input to low and then back to high.

Displaying the Result of the Half Adder in the Seven-Segment Display   The third part of the lab involves displaying the result of the adder on the seven-segment display. Figure 3.19 shows the basic connection for the TTL option: Between the common-anode display and the 74LS47.

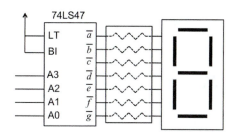

Figure 3.19. Diagram of the wiring of the seven-segment display and its driver, the 74LS47.

3.11.2 The Arithmetic Logical Unit

Required equipment: Logic board.
Required components: 74LS00 (or 74HCTS00), 74LS10 (or 74HCT10), 74LS47 (or 74HCT4511), and 74LS181.
   This lab is an exercise on the design of combinational logic circuits. It has all the main elements of the design and implementation. It can be divided into two lab periods.

Part I: Functional The first part of the exercise consists of verifying that the ALU operates as specified in Table 3.6. We do this by wiring *all* the inputs to switches and all the outputs to LEDs. We continue this verification with a more sophisticated output: The seven-segment display. This requires including its driver, the 74LS47 decoder (see previous lab project). Refer to the data sheet of the 74LS181 (not included here) for a diagram of pin connections and functions. Try a few operations with the ALU that give a result between 0 and 9. They should display nicely on the seven-segment display.

Part II: Customizing the ALU This next part involves an engineering task: Wiring the ALU so that it uses a reduced number of operations. The lab then has the following sections.

1. Solving Problem 14 provides the design of the circuitry for this lab.

2. Wire the circuit according to your solution of Problem 14. Before wiring anything, verify with the instructor that your circuit is correct (he or she must have the answers). The wiring that was performed in the first section has to be modified only slightly. Do not touch the inputs, A_is and B_is, and the outputs, F_is. Wire your decoder circuits, connecting x, y, and z to switches in the logic board. Connect the outputs of your decoder circuits to inputs S_3, S_2, S_1, S_0, M, and C_n of the ALU. Verification is, of course, a must here: For a given combination of variables A and B, check that all operations are performed successfully.

3.11.3 Practicum Test

1. Implement the circuit for the function $F = xy + x'y'$ using NAND gates.

2. Wire the output of the circuit to the seven-segment display so that segments f and b light up when $F = 1$.

3. Make a full circuit diagram of your wiring.

CHAPTER 4

ADVANCED COMBINATIONAL DEVICES

Contents

4.1 PRAGMATIC DESIGNING

When we have five or more variables, combinational design using Karnaugh maps becomes confusing and hard to do by hand. To improve efficiency, we have to resort to computational methods to solve the problem. Another option is the McCluskey-Quine method of prime implicants, which is a version of the Karnaugh map method that runs more efficiently in a computer. Such treatments are not in the scope of this course, so we do not discuss them here. Instead, we discuss how to implement circuits using clever combinations of general-purpose circuits.

As an example of doing tasks without the K-map method, let us consider the circuit that determines whether one binary number is greater than another one. Suppose that we have two 4-bit binary numbers, $\mathbf{A} = A_3A_2A_1A_0$ and $\mathbf{B} = B_3B_2B_1B_0$, and suppose that we want to generate a function G such that $G = 1$ when $\mathbf{A} > \mathbf{B}$ and $G = 0$ otherwise. Forget about writing a truth table: It involves $2^8 = 256$ combinations! Let us solve this problem in a more clever way.

Consider what we do in our heads when we compare two numbers. What do we do when we inspect two numbers to see which one is greater? We look at the most significant digit of each number: The greater of the two digits determines which number is greater. For example, if we compare two decimal numbers, 5,637 and 2,401, we immediately look at the thousands position and determine that the first number is greater because it has a 5 in that position and the other number has a 2. We would do something similar in binary. If the numbers were 1011_2 and 0110_2, we would determine that the first number is greater than the second one by comparing their MSB: It is a 1 for the first and a 0 for the second. We can state this possibility the following way: $\mathbf{A} > \mathbf{B}$ if $A_3 = 1$ and $B_3 = 0$, or if $A_3B_3' = 1$. That is just one possibility. What if $A_3 = B_3$, as in the comparison of 1101_2 and 1010_2, or 0011_2 and 0111_2? In this case we compare the next two digits. We know that $A_3 = B_3$ when either both are 0s or 1s, or when $A_3 \odot B_3 = 1$, so for the next possibility the condition is $A_3 = B_3$ AND $A_2 > B_2$, or $(A_3 \odot B_3)A_2B_2' = 1$. If you follow this rationale for our four-digit numbers, you come up with a function that involves OR-ing the four possibilities that determine that $\mathbf{A} > \mathbf{B}$:

$$
\begin{aligned}
G = {} & A_3 \cdot B_3' + (A_3 \odot B_3) \cdot (A_2 \cdot B_2') \\
& + (A_3 \odot B_3) \cdot (A_2 \odot B_2) \cdot (A_1 \cdot B_1') \\
& + (A_3 \odot B_3) \cdot (A_2 \odot B_2) \cdot (A_1 \odot B_1) \cdot (A_0 \cdot B_0')
\end{aligned}
$$

We can read this function the following way: $\mathbf{A} > \mathbf{B}$ when $(A_3 > B_3)$ or $(A_3 = B_3$ and $A_2 > B_2)$ or $(A_3 = B_3$ and $A_2 = B_2$ and $A_1 > B_1)$ or $(A_3 = B_3$ and $A_2 = B_2$ and $A_1 = B_1$ and $A_0 > B_0)$.

In reality, we do not need to wire all these gates to perform this function because the 74LS85 4-bit magnitude comparator does the job for us. However, the above example gives the spirit of pragmatic designing: We just do whatever is necessary to get the task done. In problem solving, it is useful to revisit how we solve problems similar to the ones we are facing. Recalling how we compare numbers mentally gave us a clue on how to proceed. Next we discuss other important modular circuits that we can use for many applications.

■ **EXERCISE 4.1**

We have two 3-bit numbers, $\mathbf{A} = A_2A_1A_0$ and $\mathbf{B} = B_2B_1B_0$. Each bit is represented by a digital input line. Find a function G that is a 1 when $\mathbf{A} \neq \mathbf{B}$.

4.2   ADDERS

A full adder is a circuit that adds two binary operands and an input carry. It gives rise to two types of outputs: a sum bit and a carry bit. The input and output carry bits are included to allow the possibility of connecting several adders in tandem. Shown in Figure 4.1 is the block diagram of a 1-bit full adder. The truth table for this adder is given in Table 4.1.

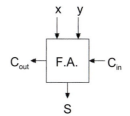

Figure 4.1.   Block diagram of a full adder.

Table 4.1.   Truth Table of a 1-Bit Full Adder

x	y	Cin	Cout	S
0	0	0	0	0
0	0	1	0	1
0	1	0	0	1
0	1	1	1	0
1	0	0	0	1
1	0	1	1	0
1	1	0	1	0
1	1	1	1	1

If we wanted to implement a 4-bit adder, we would need to account for what goes on in the following operation:

$$
\begin{array}{cccccc}
 & & & & & \text{Cin} \\
 & & A_3 & A_2 & A_1 & A_0 \\
+ & & B_3 & B_2 & B_1 & B_0 \\
\hline
\text{Cout} & S_3 & S_2 & S_1 & S_0 \\
\end{array}
$$

We can implement this by either solving the nine-variable Karnaugh map problem or, more pragmatically, by using four 1-bit full adders in tandem, as shown in Figure 4.2. The main feature of the tandem connection is that the carry-out of a given stage is connected to the carry-in of the next stage. This is simple and pragmatic. The only inconvenience is that it is a "ripple adder." That is, not all the sum bits reach their correct values at the same time. That is because one stage depends on the carry of the previous stage. If each adder has two levels of gates, then it takes two gate delays per digit. In our case, that is eight delays total. We can generate an interesting variation of the circuit that reduces the delays, which consists of adding a "look ahead carry" circuit. Because we do not want to get too distracted, we leave this alternative for one of the problems.

Table 4.2 lists a few ICs with packaged 4-bit adders. Using the same prescription, this chip can be connected in tandem with others if we want an adder with more than 4 bits. The table also shows several multi-function ICs, such as the arithmetic logical unit, or ALU.

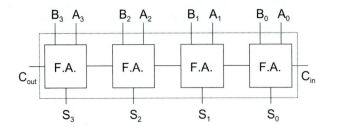

Figure 4.2.   Four-bit full adder made of four 1-bit adders.

Table 4.2.   Useful Adder ICs

IC #	Description
83, 283	4-bit binary adders
181[†]	4-bit arithmetic logic unit: 48 functions
184	BCD-to-binary converter
185	Binary-to-BCD converter
381, 382	4-bit arithmetic/logic: plus, minus, AND, OR

[†]Only in LS

4.3   DECODERS

A decoder is a straightforward device. It has n inputs and 2^n outputs. Each output is a minterm of the corresponding inputs. For example, a 2×4 decoder has the truth table shown in Table 4.3.

Table 4.3.   Truth Table of a Decoder

x	y	D_0	D_1	D_2	D_3
0	0	1	0	0	0
0	1	0	1	0	0
1	0	0	0	1	0
1	1	0	0	0	1

You can easily see that the outputs are $D_0 = x'y'$, $D_1 = x'y$, $D_2 = xy'$, and $D_3 = xy$. Decoders with a higher number of inputs follow the same design. This circuit has several interesting applications. One consists of implementing Boolean functions by just feeding the corresponding minterm outputs of the decoder to an OR gate (try the next exercise).

■ **EXERCISE 4.2**

Implement the function $F = x'y'z + x'yz + xy'z + xyz'$ *using a 3×8 decoder.*

Decoders can be used to make permanent memories, such as a read-only memory (ROM). Figure 4.3 shows an example of a 4×4 programmable read-only memory (PROM).

The select inputs of the decoder form a binary number A_1A_0 that represents the address of the memory location (four in total). When a particular address has been selected, such as 00, the corresponding output (D_0) of the decoder is a 1 and the others are zero. The output lines of the decoder are connected to four OR gates. These gates constitute the data outputs of the PROM, which represent the binary number $O_3O_2O_1O_0$. The connections to the OR gates are mediated by fuses. When programming the PROM, a different circuit not shown here blows fuses to break the connections to the proper OR gates.

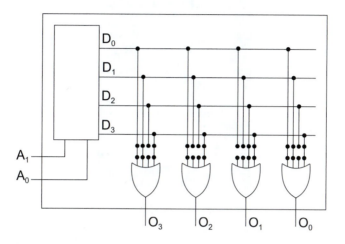

Figure 4.3. Diagram of a programmable read-only memory (PROM).

Table 4.4 contains part of the programming information of the PROM described earlier. The "stored" data for the address 00 is listed on the first row. For the output to be 1101 when $A_1A_0 = 00$, we blow the fuse that connects D_0 and O_1 (see the non connection in the figure) and leave the other fuses connected to D_0 intact. This way, when $D_0 = 1$, the lines O_3, O_2, and O_0 are set to 1, while O_1 is set to zero.

Table 4.4. Programming of the PROM of Figure 4.3

A_1	A_0	O_3	O_2	O_1	O_0
0	0	1	1	0	1
0	1				
1	0				
1	1				

■ **EXERCISE 4.3**

Fill in the remainder of Table 4.4 according to the setting of the fuses of the PROM in Figure 4.3.

After the PROM chip is programmed, the information is stored forever. In other types of ROM, the fuse is replaced by an electronic switch. The switch (a metal-oxide-semiconductor field effect transistor, or MOSFET) is controlled by charge stored in the "gate" input of the device. The charge can be removed by shining UV light on the chip. The chip can thus be reprogrammed. This is an erasable PROM, or EPROM. Most processors have these devices. You can recognize them because they have a gold sticker on top of them. Removing the sticker and shining UV light erases the programming of the EPROM. Figure 4.4 shows an EPROM without its sticker.

Figure 4.4.   View of an EPROM. The bare circuit is visible through the window. By shining ultraviolet light through the window, we can remove the connections made via stored charges in electronic switches.

Decoders are also used for memory addressing. Each output of the decoder is connected to a memory location. When the specified n-bit address is placed on the input lines of the decoder, only the line of the corresponding memory location is asserted. The other $2^n - 1$ lines are not asserted.

Suppose we want to connect one of four devices (for example, printers) to the same host. Instead of using a hardware switch, we could use a digital switch. The outputs of a 2×4 decoder could be connected to an enabling input of the printer, and its two inputs would specify the code of the device that gets enabled. This may seem unnecessary for only four devices; we could use four lines, one for each device. However, if we had more devices, say 1,000, we could condense a dedicated line for each device by specifying ten lines ($2^{10} = 1,024$) with a decoder.

The principle of decoding is manifestly used in the new alternative to the serial port: the USB port (which stands for universal serial bus). The serial port of a computer can be connected to only one device. Some PCs have more than one serial port so that they can connect to several devices. The USB system changed all this. In the USB protocol, each data package has a 7-bit sequence that identifies the device that is sending or receiving data through the serial port. This way, 128 devices can be connected to the same serial line. The difference with the decoders described earlier is that the decoding information goes along with the data itself; decoding is performed at the software level.

The situation described earlier with a hardware-based decoder enabling one of several devices does not cover all the possibilities: Under this design, one device is always on. That is, we never have all the devices disabled. We can modify our decoder slightly to fix this. We can add an "enable" input so that when it is true, the decoder works normally, asserting the selected output, but when the enable input is false, all the outputs are disabled. This function is represented by the truth table in Table 4.5.

Table 4.5. Truth Table of a Decoder with Enable

E	A_1	A_0	D_0	D_1	D_2	D_3
1	0	0	1	0	0	0
1	0	1	0	1	0	0
1	1	0	0	0	1	0
1	1	1	0	0	0	1
0	×	×	0	0	0	0

As before, the symbol "×" represents any value (0 or 1). When the enable is 0, the outputs are all 0, regardless of A_1 and A_0. Figure 4.5 shows the wiring of a decoder with enable.

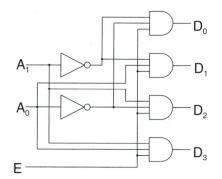

Figure 4.5. Circuit diagram for a decoder with enable.

So far, we have been working with *high-active* logic. That is, when a certain action is enabled, the output is a logic 1. Later you will see that, for reasons related to hardware and power consumption, sometimes it is better to use a *low-active* system, where devices or functions are enabled with a logic 0. For example, a decoder with low-active enable and low-active outputs has the truth table shown in Table 4.6.

Table 4.6. Truth Table of a Decoder with Low-Active Enable and Low-Active Outputs

$\overline{E}$	A_1	A_0	$\overline{D_0}$	$\overline{D_1}$	$\overline{D_2}$	$\overline{D_3}$
0	0	0	0	1	1	1
0	0	1	1	0	1	1
0	1	0	1	1	0	1
0	1	1	1	1	1	0
1	×	×	1	1	1	1

Low-active inputs or outputs are denoted with a bar on top. In a circuit diagram the low-active inputs or outputs can be alternatively denoted with a circle next to the connection on the chip but labeled *without* the bar on top, as shown in Figure 4.6. You can use either format, but not both.

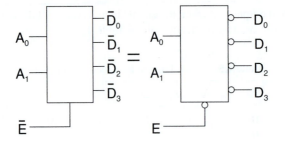

Figure 4.6.   Two ways to represent a decoder with low-active inputs and outputs.

This circuit has important applications in memory addressing. A decoder with enable can be used to decouple addressing a memory unit from actually enabling it. This way, we could be selecting a memory location by selecting the values of the address lines. This channels the enable line to the proper memory. Then when the enable line is asserted, the communication can proceed. With this architecture, memory access occurs in two steps: Setting the address first and then communicating. This is how processors communicate with memory inside computers, as shown in the next chapter.

■ **EXERCISE 4.4**

We can also use the enable line for a different purpose: To expand the decoding space. Design a circuit that uses two 2×4 decoders with enable to make a 3×8 decoder. The circuit should have three lines as inputs: the address lines A_2, A_1, and A_0; and eight lines as outputs: the data lines D_0, D_1, D_2, D_3, D_4, D_5, D_6, and D_7.

4.4   DEMULTIPLEXERS

We can look at the decoder with enable in a different way. So far, we use the enable just to turn on the selected device. In the truth tables of the decoders with enable (high-active or low-active), you will notice that the selected output has the same value as the enable. The other outputs are set to the inactive state (0 for high-active and 1 for low-active). Therefore, we could say that the x and y inputs of the decoder select the output to which the enable input is going to be connected. This type of circuit is also known as a *demultiplexer* (or DMUX; Figure 4.7). A list of decoder/demultiplexer ICs is given in Table 4.7.

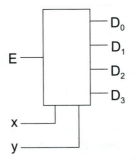

Figure 4.7.   Block diagram of a demultiplexer.

Table 4.7 lists decoder and demultiplexer ICs.

Table 4.7.   Useful Decoder/Demultiplexer ICs

IC #	Description
139, 155	2 to 4 decoder/demultiplexer
138	3 to 8 decoder/demultiplexer
154	4 to 16 decoder/demultiplexer
42	4 to BCD decoder/demultiplexer
47,49,[†]4511[‡]	BCD to seven-segment decoder

[†]Only in TTL, [‡] only in CMOS

4.5 ENCODERS

An encoder does the opposite of the decoder. It has 2^n inputs and n outputs. Table 4.8 shows the truth table of a 4×2 encoder.

Table 4.8.   Truth Table for an Encoder

I_0	I_1	I_2	I_3	x	y
1	0	0	0	0	0
0	1	0	0	0	1
0	0	1	0	1	0
0	0	0	1	1	1

The outputs (x, y) give the code of the input line that has been asserted. This circuit has applications with keypads, where the output variables give the code of the pressed key. Each push-button of the keypad is a switch that can be wired to specify a 1 or a 0. We can also expand this circuit so that it has an enable, as shown in Table 4.9.

Table 4.9.   Truth Table of a Priority Encoder

E	I_0	I_1	I_2	I_3	x	y
1	1	0	0	0	0	0
1	×	1	0	0	0	1
1	×	×	1	0	1	0
1	×	×	×	1	1	1
0	×	×	×	×	0	0

We can define an encoder with enable with low-active inputs and low-active outputs, as done for the decoder. Other types of encoders, called priority encoders, prioritize the code when two keys are pressed simultaneously (see Problem 10 in Chapter 3). The circuit of Figure 4.8 shows a circuit that uses two 74LS148 8-to-1 priority encoders to decode the 10 number keys of a keypad. Notice that some of the inputs of one of the ICs are not used. The entire 16 inputs could be used for other keys. Modern computers now have more sophisticated encoders that produce the 7-bit ASCII or the 8-bit HEX code of the key being pressed.

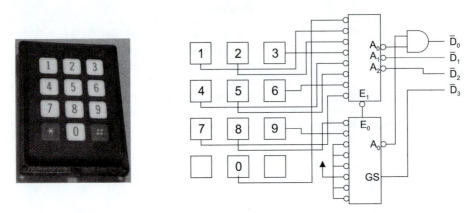

Figure 4.8.   Photo of a keypad and wiring diagram of the driving circuitry using a 74LS148 priority encoder.

4.6 MULTIPLEXERS

A demultiplexer can be seen as a switch: The select lines determine the output to which the input line gets connected. Such a switch works, but in only one direction. A *multiplexer* has the same function as a demultiplexer, but in the opposite direction. That is, the n select lines determine which of the 2^n input lines gets connected to the output line.

Figure 4.9 shows the schematic of a multiplexer (or MUX). We can "guess" that the inside of the multiplexer must look like the circuit of Figure 4.10. We can write a condensed truth table for this operation, as shown in Table 4.10.

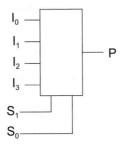

Figure 4.9.   Block diagram of a multiplexer.

Multiplexers can also be used to generate Boolean functions. The select lines of the multiplexer become the input variables of the function. We then set the line inputs of the multiplexer to be 1s and 0s, according to a truth table. Consider the function F represented by the truth table of Table 4.11.

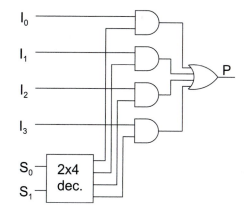

Figure 4.10.   Diagram showing how a multiplexer can be wired.

Table 4.10.   Truth Table of a Multiplexer

S_1	S_0	P
0	0	I_0
0	1	I_1
1	0	I_2
1	1	I_3

Table 4.11.   Truth Table of an Example

x	y	z	F
0	0	0	1
0	0	1	0
0	1	0	1
0	1	1	0
1	0	0	1
1	0	1	0
1	1	0	1
1	1	1	1

With an 8×1 multiplexer, we can implement it as shown in the circuit in the left side of Figure 4.11. The same function can be implemented with a 4×1 multiplexer. Notice that we use the y and z inputs as the select lines of the multiplexer. If you inspect the truth table, you will see that combinations 00, 01, and 10 of yz have the respective values of 1, 0, and 1 regardless of the value of x. Therefore, we wire the corresponding inputs to the multiplexer directly to be 1 or 0. The combination $yz = 11$ is 0 when $x = 0$, and 1 when $x = 1$, so we wire that multiplexer input to x. Table 4.12 lists useful encoder and multiplexer ICs.

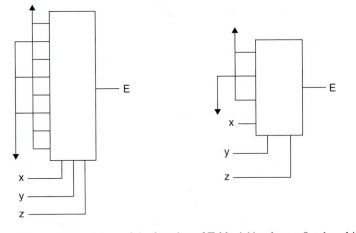

Figure 4.11.   Two possible wirings of the function of Table 4.11 using an 8×1 multiplexer (left) and a 4×1 multiplexer (right).

Table 4.12.   Useful Encoder/Multiplexer ICs

IC #	Description
148	8- to 3-line priority encoder
147	BCD 4- to 10-line priority encoder
157, 158	2×1 multiplexer
153	4×1 multiplexer
253	4×1 multiplexer with tristate output
151, 152	8×1 multiplexer
251	8×1 multiplexer with tristate output
150	16×1 multiplexer
250	16×1 multiplexer with tristate output

■ **EXERCISE 4.5**

Implement the function $F = y + x'y'$ *using a* 4×1 *multiplexer and a* 2×1 *multiplexer.*

The combination multiplexer-demultiplexer plays a vital role in telephone communications. This is shown schematically in Figure 4.12. Our phone conversations get sampled at

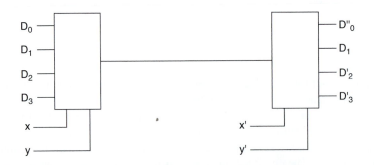

Figure 4.12.   Multiplexer-demultiplexer combination for use in communications.

a rate of 8 kHz and digitized into 8-bit numbers. The numbers are sent into big telephone multiplexers that send them through a telephone line or optical fiber to their destinations, where they are demultiplexed toward the other end user. During the time between samplings, data from other conversations are sent through the same line or fiber, but later demultiplexed toward their proper destinations.

4.7 PROBLEMS

1. We have two 4-bit numbers $\mathbf{A} = A_3A_2A_1A_0$ and $\mathbf{B} = B_3B_2B_1B_0$. Find a circuit diagram that generates a 1 when $\mathbf{A}$ is greater than or equal to $\mathbf{B}$.

2. In a full adder, each digit has three inputs (A_i, B_i, and C_{ini}) and two outputs (S_i and C_{outi}).

   (a) Using the standard logic minimization procedure, find Boolean expressions for S_i and C_{outi}.

   (b) What is the maximum number of gate delays in arriving at the two outputs?

   (c) Now consider a 3-bit full adder. How many gate delays are involved in getting the adder to settle into its final value? Show your work.

   (d) We want to shorten the time by adding "look ahead" carry circuits. A carry $C_{outi} = 1$ is generated if either: both A_i and B_i are 1, OR if A_i or B_i (or both) are 1 AND C_{ini} is 1. State this condition in a Boolean expression for each digit i, where $i = 1, 2, 3$. Let us also call $C_{in1} = C_0$, $C_{out1} = C_{in2} = C_1$, $C_{out2} = C_{in3} = C_2$, and $C_{out3} = C_3$.

3. We have two 4-bit lines representing two BCD variables. We want to design a 4-bit BCD adder. That is, the output of the circuit must be the BCD number that results from adding the two BCD numbers. We have at our disposal one or more 4-bit binary adders. Using any gates that you need and the binary adder(s), design the circuit that will do the job. *Hint*: Start by looking at the output of the 4-bit binary adder that has the two BCD variables as inputs, and compare that to the output that BCD adder should have.

4. Write down the gate-level circuit of the decoder with low-active enable and low-active outputs.

5. How would you make a 3×8 decoder using two 2×4 decoders with enable (all high-active)? (*Hint*: You can use the third input to enable/disable the decoders.)

6. Draw a circuit diagram of a PROM similar to Figure 4.3 that follows Table 4.13 with the requirements given here.

   (a) Use a decoder.

   (b) Use a low-active input enable line such that when enable is true the PROM works as planned, but when false, all the outputs of the PROM are a logic 0.

   (c) Use a low-active enable so that, when false, the outputs are in a high-impedance state.

7. Design a 1×8 demultiplexer using two 1×4 demultiplexers (74LS139). Make a diagram.

8. Use the data sheet of the 74LS148 to explain the role that E_I, E_0, and GS play in the circuit of Figure 4.8.

Table 4.13. Programming of the PROM for Problem 6

A_1	A_0	O_3	O_2	O_1	O_0
0	0	1	1	0	1
0	1	0	1	1	0
1	0	0	1	0	0
1	1	1	1	1	1

9. Implement the function $F = x'y'z + x'yz + xy'z + xyz'$ using:

   (a) A decoder
   (b) An 8×1 multiplexer
   (c) A 4×1 multiplexer

10. A full subtractor subtracts two numbers, such as $X - Y = Z$, where X is the minuend, Y is the subtrahend, and Z is the difference.

    (a) Check your binary subtracting skills by verifying that $10011011 - 01011101 = 00111110$.

    (b) A 1-bit full subtractor has as inputs the minuend x_i, the subtrahend y_i, and a borrow-in B_{in}. A borrow-in arises when the operation for the previous least significant digit needs to borrow. Its outputs are the difference z_i and borrow-out B_{out}. The borrow out is generated when the minuend has to borrow. Write the truth table of a 1-bit full subtractor.

    (c) Implement circuits for z_i and B_{out} using:

        i. An 8×1 multiplexer.
        ii. A 4×1 multiplexer.

4.8 LAB PROJECTS

This section presents ideas for lab activities.

4.8.1 Multiplexing

Required equipment: Logic board. Pre-wired 74LS90 decade counter and 2-bit up binary counter. Two square-wave function/pulse generators working at 1 Hz and another one at 0–10 kHz.

Required components: 4×1 multiplexer and demultiplexer ICs, 74LS47, and seven-segment display.

Multiplexing   Select 4×1 multiplexer and demultiplexer ICs. Draw a circuit diagram, and late-wire a circuit that has the following specifications:

1. The data inputs of the 4×1 multiplexer are connected to the outputs of the decade counter and to LEDs.

2. The clock input of the counter is connected to the 1 Hz square wave.

3. The output of the multiplexer is wired to an LED display.

4. The select inputs of the multiplexer are wired to digital I/O switches.

Verify that the multiplexer operates as expected in this circuit.

Demultiplexing   Wire a 4 × 1 demultiplexer with the following specifications:

1. The output of the demultiplexer must be wired to LEDs.

2. Wire the select inputs of the demultiplexer so that the outputs of the demultiplexer are correlated to the inputs of the multiplexer.

3. Wire the 2-bit counter to the select inputs of the multiplexer and demultiplexer so that the seven-segment display shows the output of the decade counter.

CHAPTER 5

SEQUENTIAL LOGIC

Contents

So far, we have discussed ways to combine digital signals to obtain a final outcome according to a set of conditions (for example, a truth table). Notice that the signals are all stable. Once the input variables are set, and after a short transient time, the outputs are all set and stable.

Something seems to be missing: a time variation. What if we want to generate a *sequence* of digital outputs according to some prescription? After all, processors do that when they run a program. Sequential logic treats the design of circuits that are time dependent. With the aid of new logic devices we can make new designs that can be wired to produce any sequence of outputs that we desire. This dynamic aspect of digital logic is enriching and greatly enhances our digital bag of tricks. After you read this chapter you can dare to create your own processor from scratch.

5.1   DEFINITIONS

Sequential logic is substantially different from combinational logic. It consists of the logic of circuits whose outputs depend on their value at previous times. One example of this is a counter: The present value of the count depends on its previous value. In sequential logic, we analyze circuits that are dynamic rather than static.

Since we are dealing with time-dependent logic signals, we need to cover a few definitions. Three types of time-varying logic devices exist: astable, monostable, and bistable. Astable devices are always changing state and have no stable state. A clock is an example of an astable device, always going "tick-tock." In a computer, a clock generates a signal that is constantly changing between 0 and 1, with the transitions defining the times at which events occur. Monostable signals are those that have only one stable state. Upon a certain stimulus, they momentarily change state but return to their stable state after some finite time. A television remote control is a monostable device: It sends a signal upon a stimulus when a push-button is pushed momentarily, and returns to its stable state after a certain predetermined time. A light switch is an example of a bistable mechanical device because it has two stable states: on and off. In a computer, a memory unit is bistable because it can have two stable states: 0 and 1.

Another important concept in sequential logic is the propagation delay, mentioned in Chapter 2. This is the time that it takes for a transition to propagate through the gate. The time is typically 7 ns for a gate in the 74HC00 IC, which is the digital gate with the simplest circuit. For a more elaborate gate, such as the D-flip-flop in the 74HC74 IC (see the next section), the delay is 15 ns. In general, the propagation delay in going from low to high differs slightly from the delay in going from high to low.

We can use these delays to our advantage. With them, we can create transient effects that serve a useful purpose. The circuit of Figure 5.1 shows how we take advantage of the propagation delay to make a positive-edge detector. This monostable circuit generates a pulse every time the input changes from low to high.

Suppose that the input line is a 0. The output of the NOT gate is a 1. Therefore, in the steady state (when the input does not change), the output of the AND gate is a 0. The same is true if the input is a 1: The output is also a 0. However, this is the case only in the steady state. In a transient state, the output may not be 0. Suppose that the input goes from 0 to 1. The propagation of the transition through the NOT gate delays the arrival of the transition to the input of the AND gate. During this propagation time, the inputs of the AND gate are temporarily equal. The AND gate will therefore have an output of 1 during this time. Of course, these cases cease when the high-to-low transition emerges from the NOT gate. The end result is a short positive pulse emerging from the output of the circuit every time there is a positive-edge transition in the input.

If we want to implement the circuit of Figure 5.1 we also have to worry about a few details. The propagation delay for a NOT gate is 7 ns, but this is also of the order of the

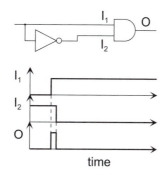

Figure 5.1. Positive-edge detector.

propagation delay of the AND gate. To keep the input of the AND gate from changing faster than its propagation delay, we must add more than just a NOT gate. Perhaps we could make it an odd number of NOT gates, such as five, so that the delay is about 35 ns. Although propagation delays are real, we cannot rely too much on the actual value of the time delays because these may vary with temperature and from gate to gate. In the LS family, there is the 74LS31 IC, which contains several gates with different propagation delays (see data sheets).

Another issue is that a few nanoseconds is a short time, involving effective frequencies of the order of 100 MHz. At these frequencies, the wiring of the circuit is important because the transient response of the other circuit elements (wires behaving as capacitors and inductors) becomes important; instead of having a nice, clean transition, you get some wiggle that looks like anything but a square pulse. In this regime, you must also use short wires. If the change is of the order of the time it takes the signal to propagate through the wire (a fraction of the speed of light, which is 1 foot per nanosecond), then the propagation of a signal through it looks more like a wave propagating through the wire. When the signal reaches the end of the wire, some of it will be reflected back. This is called ringing. Welcome to the real world!

■ **EXERCISE 5.1**

The circuit of the previous example works for positive-edge transitions. Design one that works for negative-edge transitions.

5.2 FLIP-FLOPS

A flip-flop is a bistable device. It takes advantage of the inherent propagation delay of gates to "latch" binary information. Flip-flops are the basis of registers and memory. The RS flip-flop is the simplest type of bistable device. Let us understand how it works with the following exercise.

Consider the system shown in Figure 5.2. The initial values of the lines are $S = 0, R = 0$, $Q = 1$, and $Q' = 0$. We do this example in steps. First, let us find the value of Q and Q' when we make $R = 1$. Note that one of the inputs of each NOR gate relies on the output of the other gate, so we have to make sure that the final outputs are stable. When $R = 1$, it switches Q to 0. Because Q is connected to the input of the lower gate, it changes Q' to 1. When we feed the new value of Q' to the top NOR gate, we see that it does not change the value of Q. Therefore, making $R = 1$ changes the outputs of Q and Q' to new values: 0 and 1,

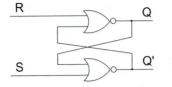

Figure 5.2. RS flip-flop with NOR gates.

respectively. Suppose that we change R back to zero. Notice that this does not change the value of Q and, therefore, Q'. Now you can go through the following exercise.

■ **EXERCISE 5.2**

Continue where we left off in the previous circuit for the changes mentioned and fill in Table 5.1.

1. *Find the value of Q and Q' when we make $R=0$.*
2. *Find the value of Q and Q' when we make $S=1$.*
3. *Find the value of Q and Q' when we make $S=0$.*
4. *Find the value of Q and Q' when we make $R=1$.*
5. *Find the value of Q and Q' when we make $R=0$ and $S=1$.*
6. *Generalize the function of the circuit but without considering the $S=1$, $R=1$ state.*

Table 5.1. Truth Table for Sequential Inputs to the RS Flip-Flop

S	R	Q	Q'
0	0	1	0
0	1	0	1
0	0		
1	0		
0	0		
0	1		
1	0		
0	0		

From this exercise you must conclude that the circuit is stable for three sets of values of its inputs S and R: 00, 01, and 10. For 11 the flip-flop is unstable. To see this, continue Table 5.1, with the next state being 11. You will find a stable output with both Q and Q' zero. Now continue, but switch *both* inputs to zero simultaneously—that is, propagate the logic signals at the same time. You will see that the output keeps changing without end. If there is a temporal imbalance in the two sides the output will settle to 10 or 01, but we will not be able to predict it. For this reason, we never use the 11 input state in the SR flip-flop.

To summarize, three inputs of the SR have a definite function. When $S=0$ and $R=0$, the outputs do not change. It can be said that in this state the flip-flop is "remembering." When $S=1$ and $R=0$, the outputs Q and Q' are set to 1 and 0, respectively. This is called the "set" state. When $S=0$ and $R=1$, Q and Q' are 0 and 1, respectively, so this is called the "reset" state. Notice that Q and Q' are the opposite of each other. This is an added bonus of the circuit that we use to our advantage when designing circuits. The output Q can be thought of as a memory cell. Table 5.2 gives a functional truth table for the RS flip-flop. In the table, Q_{t-1} is the value of Q_t in the previous instant (the state of Q before $S = 0$ and $R = 0$). Again, the $S = 1$, $R = 1$ state is indeterminate and should be avoided.

Table 5.2. Condensed Truth Table for the RS Flip-Flop

S	R	Q_t	Action
0	0	Q_{t-1}	No change
1	0	1	Set
0	1	0	Reset
1	1	?	Indeterminate, unstable

5.3 D FLIP-FLOP

We can avoid the $S = 1$, $R = 1$ hazard by forcing the two inputs to be opposite of each other. In the circuit of Figure 5.3, R is forced to be NOT-S, so S and R are always opposites of each other. The input D then controls the state of Q and Q'. This is called the D flip-flop. One shortcoming of this circuit is that the output is always setting or resetting, responding constantly to the current value of D. This circuit is not very useful, but we can turn it into a useful one by introducing an enabling input, as shown in Figure 5.4. Table 5.3 gives the corresponding truth table.

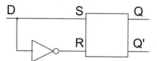

Figure 5.3. D flip-flop.

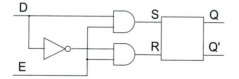

Figure 5.4. D flip-flop with enable.

Table 5.3. Condensed Truth Table of the D Flip-Flop with Enable

E	Q_t
0	Q_{t-1}
1	D

■ EXERCISE 5.3

Draw the timing diagram that corresponds to the input signals of the D flip-flop with enable shown in Figure 5.5.

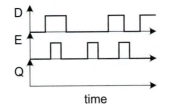

Figure 5.5. Input/output waveforms for Exercise 5.3.

As we see later, a memory cell works this way. When $E = 1$, we load the flip-flop with the value of D, and when $E = 0$, the flip-flop ignores the value of D and "remembers" its state.

■ EXERCISE 5.4

As a preview of our next section, consider the circuit of Figure 5.6. It consists of two flip-flops connected in a peculiar way. The method of triggering is called the master-slave method. If we use the timing diagram of Figure 5.3 for D and E, draw the timing diagram for X and Y. Draw your conclusions.

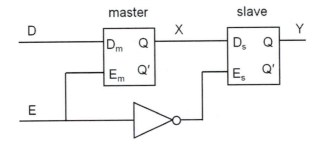

Figure 5.6. Master-slave triggering of flip-flops.

5.4 EDGE-TRIGGER

In the previous application the flip-flop outputs can change as long as $E = 1$. In some applications, we want the change to occur in a short amount of time, with the flip-flop taking a "snap shot" of the input. We can implement this by making the enable sensitive to the *change* in the enable input. The master-slave circuit of Figure 5.6 (Exercise 5.4) improves upon this: The input signal is latched at the negative transition of the clock pulse. This triggering can be improved further by having a circuit react to the change in level of the input clock and not requiring a pulse. The change can then be either low-to-high or high-to-low. This transition-sensitive enable is called *edge-trigger*. The circuit shown on page 37 implements the edge-trigger (see Problem 5).

The timing diagram of Figure 5.7 shows how a positive edge-trigger flip-flop works. The timing diagram shows that the information gets transferred from input to output only when the edge-trigger enable goes from low to high. At other times, the output remains the same. Table 5.4 gives a truth table and diagram of this type of flip-flop. Figure 5.8 shows the block diagram of the flip-flop.

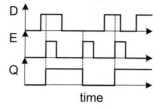

Figure 5.7.   Input/output timing diagram of a D flip-flop with edge-trigger enable.

Table 5.4.   Condensed Truth Table for D Flip-Flop with Edge-Trigger Enable

E	Q_t
×	Q_{t-1}
↑	D

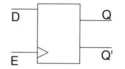

Figure 5.8.   Schematic of D flip-flop with a positive edge-trigger enable.

Notice the new labeling in the block diagram of the flip-flop: The triangular mark at the input of the enable denotes the edge-trigger capability. A circle on the outside of the edge-trigger input indicates that it is a trigger enabled by a negative edge (see the next exercise).

■ **EXERCISE 5.5**

Complete the timing diagram of Figure 5.9 for a D flip-flop with a negative edge-trigger (enabling the operation at the low-going edge of the enable input).

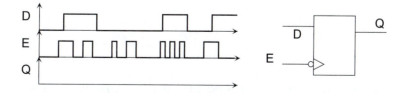

Figure 5.9.   Waveform and circuit for Exercise 5.5.

Flip-flop ICs come with two other functions that can become quite useful: clear and preset inputs. Normally, these inputs are low-active. When the clear input is set to 0, the output Q is forced to 0 (and Q' is forced to 1) regardless of the current state of the flip-flop. Conversely, when the preset input is set to 0, the flip-flop output Q is forced to be a 1 (and Q' is forced to be a 0).

A digital lock is an interesting application of flip-flops. Consider the circuit shown in Figure 5.10. The enable inputs of each flip-flop are hard-wired to four keys of a keypad. If initially all the flip-flops have been cleared (by setting the CLEAR line to 0), then only pressing the correct keys in the correct order (such as 1, then 5, then 4, then 8) will propagate the 1 from the left input to the end of the chain. A 1 after the last flip-flop could open the lock (and, for example, a door).

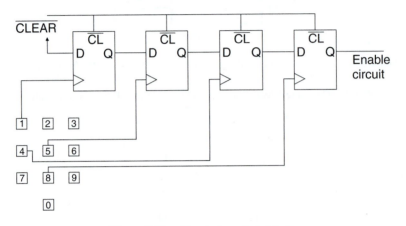

Figure 5.10. Circuit for a digital lock.

5.5 JK AND T FLIP-FLOPS

The JK flip-flop is a useful device. The first three states of the JK flip-flop are the same as those of the RS flip-flop. The fourth state, the 11 state, is now the toggle state. When the toggle state is enabled, the output changes to the opposite of what it was before the arrival of the enable trigger. Table 5.5 gives the truth table for a JK flip-flop.

Table 5.5. Truth Table of the JK Flip-Flop

J	K	Q_t	Action
0	0	Q_{t-1}	No change
1	0	1	Set
0	1	0	Reset
1	1	Q'_{t-1}	Toggle

A useful variation of the JK flip-flop is the T flip-flop. It is a JK flip-flop with its inputs tied together to a single input line, T, as shown in Figure 5.11. When $T = 0$, the output does not change, but when $T = 1$ the output flips to the opposite of what it was before. Table 5.6 shows a condensed truth table of the flip-flop. A list of flip-flop ICs is given in Table 5.7. This type of flip-flop has important applications in counters, as we see in Section 5.6.4.

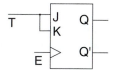

Figure 5.11.   Diagram of the T flip-flop.

Table 5.6.   Condensed Truth Table of the T Flip-Flop

T	Q_t	Action
0	Q_{t-1}	No change
1	Q'_{t-1}	Toggle

Table 5.7.   Useful Flip/Flop and Latch ICs

IC #	Description
74	D flip-flop (edge-trigger)
175	Quad D flip-flop with common clock and clear (edge-trigger)
174	Hex D flip-flop with common clock and clear (no $\overline{Q}$, edge-trigger)
75, 77	Quad D flip-flop with pulse enable (latch)
373	Octal D flip-flop with common edge-trigger clock (latch) and 3-state output
573	Octal D flip-flop with common pulse enable (latch) and 3-state output
259	8-bit addressable D latches (memory) with pulse enable
279	RS flip-flop with no enable
76,109,112	JK flip-flop (edge-trigger)

5.6   APPLICATIONS OF FLIP-FLOPS

Flip-flops have numerous applications. Here we dicuss a few important ones.

5.6.1   Latch or Register

We can use flip-flops to store data. In the circuit of Figure 5.12, two input data lines are each connected to a D flip-flop. The enable inputs are connected to a common strobe input labeled CP. When a strobe signal triggers the flip-flops, the value of the input lines D_1 and

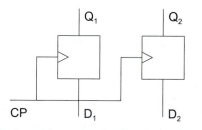

Figure 5.12.   Flip-flop registers.

D_2 gets transferred to the respective outputs of the flip-flops Q_1 and Q_2. The information in D_1 and D_2 is latched or stored. After the positive edge of the clock input is passed, the Q outputs are independent of the inputs until the next enable input arrives. This type of latch is used as a register in computer processors. Registers are general-purpose storage units for processing binary numbers and codes.

5.6.2 Frequency Divider

Flip-flops can serve as frequency dividers. See, for example, the circuit shown in Figure 5.13. When the flip-flop gets triggered, the value of the input D is transferred to the output Q. Because Q' is connected to D through a wire, the output changes when the flip-flop gets triggered. The propagation delay of the triggering action through the flip-flop ensures that, by the time the output Q' changes, the trigger edge has passed. Thus, by taking advantage of the propagation delay in the flip-flop, we can have the output change state every time the input goes from low to high (in the case of a positive edge flip-flop). We can appreciate this in the timing diagram shown in Figure 5.14. We have exaggerated the propagation delay to illustrate the effect. The propagation delay through the wire that connects Q' to D is negligible. Notice that the output completes one full period for every two periods of the input. We can also construct a frequency divider with the JK flip-flop operating in the toggle mode, as shown in Exercise 5.6.

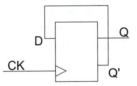

Figure 5.13.   Frequency divider with a D flip-flop.

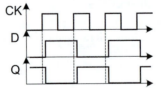

Figure 5.14.   Timing diagram of the frequency divider of Figure 5.13.

■ **EXERCISE 5.6**

The JK flip-flop in Figure 5.15 is in toggle mode. Complete the timing diagram adjacent to it. Assume that the output Q is a 0 at the start of the diagram.

5.6.3 Switch Debouncers

When we depress or release a switch, we have a short time (a few milliseconds) in which the metal contacts bounce among one another. As a result, the signal is noisy, as shown in the top trace in Figure 5.16. If we put a NOT gate after the switch, we get the waveform

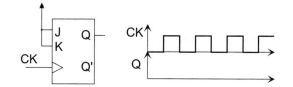

Figure 5.15.   JK flip-flop as a frequency divider.

shown in the second trace. If this signal is sent to a pulse-counting circuit, the circuit senses several pulses or transitions rather than a single transition.

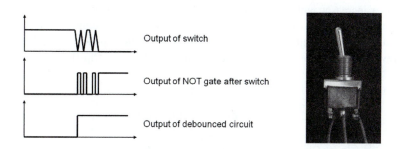

Figure 5.16.   Bouncing signal in a switch (left). View of a SPDT switch (right). It has three connections. The two positions of the switch connect the middle connection to either of the two other connections.

A solution to this problem is to *debounce* the switch. We need a specific kind of switch: a single-pole double-throw (SPDT) switch, shown in Figure 5.16. This switch has three contacts. An input to the middle contact of the switch is connected to one of two adjacent contacts.Thus, the switch has two positions. We solve the bouncing of contacts with the circuit shown in Figure 5.17. The resistors can be 1 kΩ resistors. Note that the RS latch has inverting inputs. This is the default of the NAND-based flip-flops. A latch such as the 74LS279 will work. Otherwise, wiring with discrete NAND gates works (see Problem 1). Later in this chapter, we look into two other types of debouncer circuits.

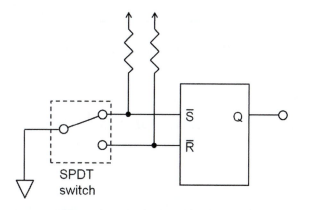

Figure 5.17.   Switch debouncer with RS flip-flop.

5.6.4 Counters

Two type of counters exist: asynchronous and synchronous. They consist of a series of flip-flops with their inputs connected in such a way that when a count pulse arrives, their outputs change according to a binary counting sequence.

Asynchronous Counters   These are counters that consist of flip-flops connected in tandem. The flip-flop of one stage generates the clocking input for the flip-flop of the next stage. For this reason, these are also known as ripple counters.

Figure 5.18 gives an example of an asynchronous counter. The circuit looks like a set of frequency dividers connected in series. When analyzing the timing diagram shown in Figure 5.19, we can see that the outputs go through states that correspond to a binary number. This number increases with every input pulse. The circuit considered here is only one of many. Notice that if we use the Q outputs of each flip-flop as the counter digits, the counter counts down. Other variations exist, so we could trigger the flip-flops with the Q' output of the previous flip-flop or we could use a negative edge-trigger flip-flop.

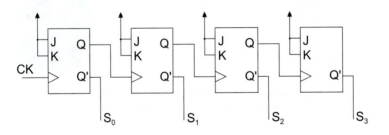

Figure 5.18.   Asynchronous 4-bit binary counter.

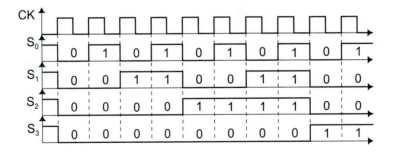

Figure 5.19.   Outputs of the circuit of Figure 5.18

■ **EXERCISE 5.7**

Design an asynchronous up-counter using JK flip-flops with negative edge-trigger. Verify its operation with a timing diagram.

■ **EXERCISE 5.8**

Complete Table 5.8, listing all the possible variations of inputs for an asynchronous counter that uses T flip-flops (JK flip-flops with inputs tied).

Table 5.8. Table for Exercise 5.8

Trigger	Trigger Comes From	Count Up	Count Down
Positive	Q	Q'	Q
Positive	Q'		
Negative	Q		
Negative	Q'		

■ **EXERCISE 5.9**

Design a 2-bit asynchronous up-counter using D flip-flops.

Asynchronous counters have shortcomings. The different digits settle to their final value only after receiving the enabling input from the previous stage. The outputs are therefore not synchronized. As a consequence, false states may be generated while the outputs are settling to a new value. For example, consider the transition from 0011_2 to 0100_2 in the counter of Figure 5.18 and 5.19. We have blown up this transition in Figure 5.20. We can see that S_0 reaches its final state a propagation delay *before* output S_1. This is because the edge that triggers flip-flop 2 is the output Q_0 of the first flip-flop. Similarly, S_2 makes the transition upon the arrival of the edge-trigger from Q_1, which occurs at the same time S_1 makes the transition. As a result, in going from 0011_2 to 0100_2 the outputs of the counter go through the following sequence of states: $0011_2 \rightarrow 0010_2 \rightarrow 0000_2 \rightarrow 0100_2$.

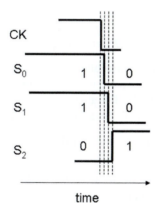

Figure 5.20. Blow-up of a particular transition in the ripple counter.

Since false states last for as long as the signal propagates through the flip-flops ($\sim25\ \mu s \times$ number of flip-flops), this may not be a factor in many situations. For example, in a digital clock read by humans, the eye does not notice false states that last 25 ns.

Asynchronous counters can be used as a "divide by 2^n" device by connecting n flip-flops in a counter configuration. The last counter puts out the divided output. If we used the clear and preset inputs of the flip-flops (discussed in Section 5.6.4), we could design a circuit that divides the clock input by any number.

Synchronous Counters and State Machines   If the counter is driving other electronic devices that rely on the synchronization of the clock inputs, we need *synchronous* counters. In a synchronous counter, all the clock inputs are connected to a master input clock signal. We must cleverly direct the outputs of the flip-flops back to their inputs, as shown next. One also relies on the inherent propagation delay of the flip-flops to make this happen.

If you look at the outputs of a counter as it is counting, as shown in Figure 5.19, you will notice that an output changes every other time that the previous lower-significant output changes (for example, S_1 changes every other time S_0 changes). Now consider the 2-bit synchronous counter shown in Figure 5.21. The first flip-flop is set to the toggle mode so that its output represents the first bit of the counter. The flip-flop representing the second bit must change every other cycle of the clock. To accomplish this, we connect the JK inputs of the second flip-flop to the Q output of the previous flip-flop.

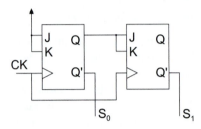

Figure 5.21.   Synchronous counter.

■ **EXERCISE 5.10**

For the circuit of Figure 5.21 make a timing diagram of the clock and the outputs Q and Q' of each flip-flop, for nine cycles of the clock. Make the propagation delays noticeably large in your diagram. Do the S outputs count up or down?

Designing synchronous counters that follow an arbitrary sequence is an interesting variation of previous designs. These are more generally known as *state machines*. We describe this type of design with an example. Let us design a modulo-6 counter. This is a counter that counts up to 5 and then resets to 0. To design this counter we need to follow these steps:

1. Select the number of bits and the type of flip-flop. In our example, we need 3 bits to specify all the counts.

2. Next we decide on the type of flip-flop. We settle for a simple one: The T flip-flop (but you can do it with others as well).

3. Then we decide the sequence of states. In our example, we need the states 000_2, 001_2, 010_2, 011_2, 100_2, and 101_2. A variation of this design is to do a different sequence when an additional input is asserted. In that case, you add that output as an extra bit and include it as part of the sequence. We do not cover this case now, but see Problem 8 for an exercise on the mentioned variation.

4. Now we fill in an elaborate table of states, shown in Table 5.9, known as an excitation table. In the first set of columns we put the "present state" of the outputs of the flip-flops. The second set of columns have the "future state" of the counter once the clock edge arrives. Here we program the counter for the consecutive states that we want it to have. Notice that we put a future state even for the unused states (110_2 and 111_2). This is important because, at power-up, the counter may land in an unused state. When this happens, we need to get it out of that state and into one of the used states, such as 000_2. If one of the unused states, such as 110_2 in this example, has 110_2 as a

future state, then if somehow the circuit lands on this state it will stay there forever and "hang up" (sound familiar?). In the example, we pick 000_2 as the next state of 110_2 so that even if the circuit ends in this unused state, it goes to a used state after the next clock cycle.

5. In the third set of columns, we enter the inputs that will make the flip-flops change to the desired values when the clock arrives.

Table 5.9. Excitation Table for T Flip-Flops in the Design of a Modulo-6 Counter

$Q_A(t)$	$Q_B(t)$	$Q_C(t)$	$Q_A(t+1)$	$Q_B(t+1)$	$Q_C(t+1)$	T_A	T_B	T_C
0	0	0	0	0	1	0	0	1
0	0	1	0	1	0	0	1	1
0	1	0	0	1	1	0	0	1
0	1	1	1	0	0	1	1	1
1	0	0	1	0	1	0	0	1
1	0	1	0	0	0	1	0	1
1	1	0	0	0	0	1	1	0
1	1	1	0	0	0	1	1	1

6. We then take the flip-flop inputs (T_A, T_B, and T_C) as functions of the flip-flop outputs ($Q_A(t)$, $Q_B(t)$, and $Q_C(t)$) and find Boolean expressions for them (using the Karnaugh map method). In our example, they are $T_A = Q_A Q_B + Q_A Q_C + Q_B Q_C$, $T_B = Q_A Q_B + Q'_A Q_C$, and $T_C = Q'_A + Q'_B + Q_C = (Q_A Q_B Q'_C)'$.

7. We finish by wiring the circuit, as shown in Figure 5.22.

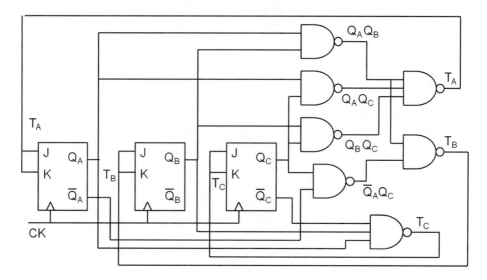

Figure 5.22. Circuit diagram of the modulo-6 synchronous counter.

We can use the previous method to step through any sequence of states. For example, we could use this to construct a pseudo-random number generator. It is "pseudo" because we decide what the "random" sequence of numbers is ahead of time, and it repeats when the cycle is completed. In Section 5.7 we present an easier way of making pseudo-random number generators with shift registers.

With this method, we could also step through any combination of numbers that we desire, for further processing. The lowest level of programming of processors and computers, "machine language" or "assembly language" instruction codes, feeds state machines to make them step through specific operations, such as memory transfers or arithmetic computations. Higher-level operations are just combinations of a core of instructions. There you have it: That is how computers work!

Counter ICs   Although we can build our own counters with individual flip-flops, a number of different counter chips are available. They come in a variety of types, with a clear input to set the count to zero, a preset input to set the count to all 1s, an up/down input that decides the counting direction, and a load input that enables you to load all the counter values with a set of inputs. The chips also come with a "carry output" line that generates a pulse every time the count goes back to zero after reaching full count, or when counting down and going from zero to full count. The latter is used for connecting counters in tandem. Table 5.10 lists a few popular 4-bit counters.

Table 5.10.   Useful Counter ICs

IC #	Type	Sync/Async	Up/Down	Trigger	Load
90	Decade	Async	No	Neg	No
92	Divide-by-12	Async	No	Neg	No
93	Binary	Async	No	Neg	No
160, 162	Decade	Sync	No	Pos	Yes
161, 163	Binary	Sync	No	Pos	Yes
168, 190, 192	Decade	Async	Yes	Pos	Yes
196	Decade	Async	No	Neg	Yes
191, 193	Binary	Async	Yes	Pos	Yes
390	Dual decade	Async	No	Neg	No
390	Dual binary	Async	No	Neg	No

Another benefit of counter ICs is that we can make "divide by N" circuits. We do this by resetting the counter when it reaches a specified value. The circuit of Figure 5.23 shows a

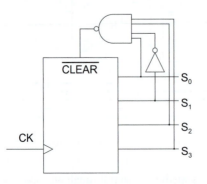

Figure 5.23.   Divide by N counter.

binary counter that resets to zero when it reaches 1101. Because the 1101 state gets switched as soon as it occurs, the counter has effectively 13 states (including zero). Notice that the clearing input is low active, one reason we use a NAND gate. Also notice that we can use the reset pulse as divide-by-13 output. Finally, if you use the circuit in this configuration, the counter must be a *synchronous* counter because a false state in an asynchronous counter may clear the counter at an unwanted time. Therefore, you must read the specifications of the counter IC that you pick.

5.7   SHIFT REGISTERS

Shift registers perform a simple but important function: They convert serial to parallel digital signals and vice versa. This is useful for communications because the parallel data in a computer can be put into serial form and sent over a single line. At the other end, the serial data is converted back to parallel for further processing. Such is the case with Internet communications via serial ports.

The diagram of Figure 5.24 shows a universal serial register. It is a simple arrangement of flip-flops. The addition of multiplexers allows us to specify whether the data is shifted to the right or to the left, or loaded in a parallel way.

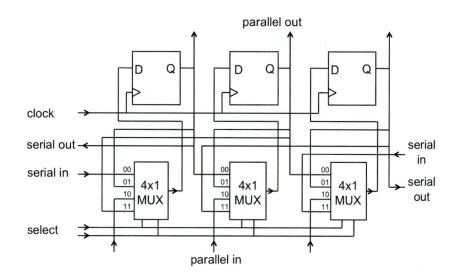

Figure 5.24.   General-purpose shift register. Multiplexer (MUX) select lines specify the following operations: $00 =$ shift right, $01 =$ no shift, $10 =$ parallel in, $11 =$ shift left.

Another use of shift registers is to generate pseudo-random numbers. These are used in cryptography, wireless communications, spread-spectrum broadcasting, and global positioning. The device that generates the pseudo-random sequence is the linear feedback shift register (LFSR). The algorithm is simple: It consists of doing successive shifts in a shift register, but with a shift-in generated from a combination of the digits of the shift register. For example, in the case of the 3-bit LFSR, the shift input is

$$S_{\text{in}} = S_0 \oplus S_1 \qquad (5.1)$$

where S_0 and S_1 are the least and next-to-least significant bits of the register. Figure 5.25 shows the wiring of a 3-bit LFSR. For 4 bits, the input shift is generated also via Equation 5.1, but for 5 bits, it is

$$S_{in} = S_0 \oplus S_2 \tag{5.2}$$

The LFSR is easy to wire with shift register ICs (see Table 5.11). The LFSR states cycle through the entire pseudo-random combination of values except zero. Therefore, the LFSR has to be set so that the initial value is not zero. If this is the case, then the shift bits are always zero and the sequence does not advance.

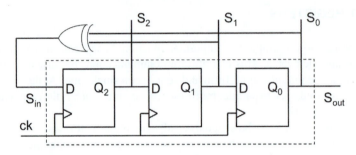

Figure 5.25.   Main design behind the 3-bit linear feedback registers for generating pseudo-random number sequences.

Table 5.11.   Useful Shift-Register ICs

IC #	# Bits	Description
95, 99, 179, 194, 195, 295	4	Parallel in, parallel out
94	4	Serial in, serial out
91	8	Serial in, serial out
198, 199	8	Parallel in, parallel out
164	8	Serial in, parallel out
165, 166	8	Parallel in, serial out
671, 672	4	Parallel in, parallel out, latched
594, 595	8	Serial in, parallel out, latched
597, 589	8	Parallel in, serial out, latched

■ **EXERCISE 5.11**

If the first state of the 3-bit LSFR is 001_2, the next state can be calculated with Equation 5.1: 100_2.

1. *Calculate the remaining states in the sequence.*

2. *Explain why we should avoid state 000_2.*

5.8   MULTIVIBRATORS

Some ICs can be used as astable clock sources. These are useful for establishing the timing pattern of a circuit. When frequency stability is required, we must resort to specialty ICs that rely on the oscillations of a repeatable quartz crystal. When stability is not a critical parameter we can use the popular LM555 timer IC, which can be made to function as an oscillator. In Chapter 10 we discuss the internal circuitry of this IC. In the labs for sequential logic, we use the LM555 timer to generate the clock signal for a counter.

A general purpose clock is used as a trigger in many applications. If the frequency of the clock is near a multiple of the 60 Hz of the line, we can get spurious signals, as 60-Hz induced noise is not averaged out fast enough. A solution to this is to generate a clock signal from the line frequency. We go over this exercise later in the analog part of the course.

If the objective is to send a single pulse, then monostable or "single-shot" circuits are necessary. Earlier in this chapter, we discussed a monostable circuit that used a NOT gate's propagation delay. However, this may be too short for many practical circuits. Table 5.12 shows some alternatives.

Table 5.12. Useful Multivibrator ICs

IC #	Description
LM555	General-Purpose Timer
121, 123, 423	Single Shot
31	Delay

5.9 MEMORY

An important component of digital computing is the capability to store data. To do this, we need the storage device, memory, and the logic mechanism for storing and retrieving the information. In this section, we discuss how memory works and how it is addressed.

Today we have several types of memory. We already covered ROM. Here we discuss the operation of random access memory (RAM). We can subdivide RAM into two types: static and dynamic. In this discussion we limit ourselves to static RAM. Flash memory is a modern form of RAM. We do not discuss other forms of data storage, such as magnetic disks and tapes, or compact disks. The technology associated with these devices is nonetheless interesting and clever. We start our discussion with the basics of a memory unit.

5.9.1 Memory Cell

A schematic of the basic memory unit is shown in Figure 5.26. It has three input lines and one output line. Two of the input lines are control lines. One of them, $\overline{\text{SEL}}$, a low-active input, enables the memory cell. The other control line, R/$\overline{\text{W}}$, specifies whether the operation is a read or a write. The two other lines are the data input (I) and data output (O) lines. The core of the unit works like a D flip-flop with enable. It is wired according to Table 5.13.

The wiring of the memory cell is shown in Figure 5.26 When $\overline{\text{SEL}}$ is 0, the memory cell is active. The R/$\overline{\text{W}}$ line determines the type of activity. If R/$\overline{\text{W}} = 0$ (a "write" into memory), the circuitry asserts the enable-input of the flip-flop to store the data (a logic level) on the I/O line. It also sets $T = 0$ on the tristate buffer to isolate the flip-flop's output from the I/O line. Because there is only one source putting data on the line, no conflict can arise. If a processor sets the I/O line with data for storage, the outputs of the memory cell should not be trying to set the same line to a different logic level. The tristate buffer provides this function that isolates the source of the data.

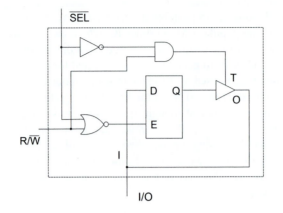

Figure 5.26.   Wiring of a memory cell.

Table 5.13.   Truth Table of the Memory Cell

Operation	$\overline{SEL}$	$R/\overline{W}$	E	T	Q	O
Write	0	0	1	0	I	H.I.
Read	0	1	0	1	Q_0	Q_0
Remember	1	×	0	0	Q_0	H.I.

When $R/\overline{W} = 1$, the memory cell is set to be "read." The enable of the flip-flop is set to 0 to disable writing into the flip-flop, and the tristate buffer is set to 1, which connects the logic output of the flip-flop to the I/O line. All the devices connected to the I/O data line must be wired so that, at any given time, only one of them puts data on the line. Memory addressing solves the apparent conflict with multiple memories. Think of the way we interact in a classroom, but with a special rule: The airwaves can be used by only one person at any given time. In terms of a class, the central processing unit would be the professor and memory would be the students. If after rambling on (hogging the airwaves as always), the professor decides to ask a question to a particular student, Kelly, then the professor proceeds by calling her by her name: "Kelly what do you think?" Thereafter, the professor waits for a response. In doing so, he reserves the air waves only for Kelly. The rest of the students remain silent while (frozen) Kelly responds. This is how memory works. It is a one-to-one communication, and never more than one device (person) uses the medium (air). Memory is hardwired so that this is guaranteed to happen, and when addressed, it always replies (professors only wish).

5.9.2   Memory ICs

Memories come in integrated circuits specified by the number of memory locations, and the number of bits per location m. The memory locations inside the chip are addressed by a set of n address lines. Thus, there are 2^n memory locations. Memory chips are normally labeled as $2^n \times m$. Figure 5.27 shows the schematic for a 4×2 memory. You can see two address lines for addressing the four memory locations. Each memory location has 2 bits, but in reality, six memory cells, two assigned to each memory address. The address lines feed into a demultiplexer, which channels the enabling select line to the corresponding memory units. Notice that pairs of memory cells get wired to the same address. Because the memory cells have tristate outputs, all I/O lines of the memory cells corresponding to the same digit position are wired together. All memory cells need the read/write control line, so they are all wired to that line.

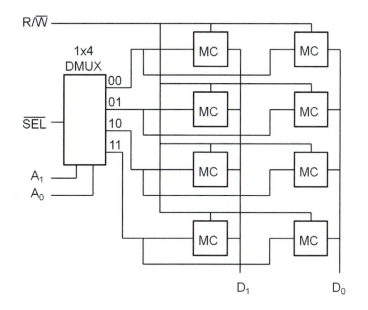

Figure 5.27.   Diagram of a 4 × 2 memory unit.

5.9.3 Memory Addressing

The address of memory in a processor is specified by a binary number. If this number has 16 bits, then the total address space is $2^{16} = 65536$. The beginning and end of this space are the addresses 0000_{16} and $FFFF_{16}$. Suppose that we have two 10-bit memory chips, and we decide to allocate their space to be contiguous, starting with the memory location 0000_{16}. We note that $1023 = 03FF_{16}$. How do we wire them? We start by discussing the address space of the chips. Table 5.14 shows the starting and ending addresses of each chip in the 16-bit space.

The addresses of the first chip are all the numbers between 0000_{16} and $03FF_{16}$ (1024 locations, including 0000_{16}). Since the space of the next chip, Chip-2, is contiguous to Chip-1, it should start at an address that is equal to the last address of Chip-1 plus 1, or 0400. The ending address of Chip-2 then is $0400_{16} + 03FF_{16} = 07FF_{16}$. Notice that all the binary addresses in Chip-1 have the following form: 0000 00XX XXXX XXXX_2, where "X" stands for either 1 or 0. Notice also that all of the binary addresses in Chip-2 have the form: 0000 01XX XXXX XXXX_2. You can see that the 11 least significant digits of the address distinguish one address from the other. Each memory location is a unique combination of 1s and 0s. Indeed, 1,024 is the total number of possible combinations of 10 binary digits. Thus a memory chip with 1,024 memory locations needs 10 address lines to uniquely specify each address within the chip. The 11th address line distinguishes the two chips.

Figure 5.28 shows the wiring of the two memory chips with capacity 1024 × 4 in a system with a 16-bit memory space. We can see in the circuit that the 10 least significant address lines A_9, A_8, A_7, A_6, A_5, A_4, A_3, A_2, A_1, and A_0 have been wired to the address inputs of both chips. Notice from the previous table that, for both chips,

Table 5.14. Address Space of the Memory Chip

	$A_{15}\ A_{14}\ A_{13}\ A_{12}$	$A_{11}\ A_{10}\ A_9\ A_8$	$A_7\ A_6\ A_5\ A_4$	$A_3\ A_2\ A_1\ A_0$	Hex
Chip-1 start	0 0 0 0	0 0 0 0	0 0 0 0	0 0 0 0	0000
	⋮	⋮	⋮	⋮	⋮
	0 0 0 0	0 0 1 0	0 1 0 0	1 1 0 0	024C
	⋮	⋮	⋮	⋮	⋮
Chip-1 end	0 0 0 0	0 0 1 1	1 1 1 1	1 1 1 1	03FF
Chip-2 start	0 0 0 0	0 1 0 0	0 0 0 0	0 0 0 0	0400
	⋮	⋮	⋮	⋮	⋮
Chip-2 end	0 0 0 0	0 1 1 1	1 1 1 1	1 1 1 1	07FF

$A_{15} = A_{14} = A_{13} = A_{12} = A_{11} = 0$. Furthermore, notice that the 11th line is $A_{10} = 0$ for all the addresses in Chip-1 and $A_{10} = 1$ for Chip-2.

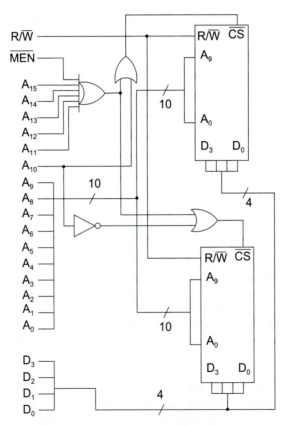

Figure 5.28. Wiring of two memory ICs.

Processors have a low-active "memory enable" ($\overline{\text{MEN}}$) line to specify a memory operation (as opposed to other operations, such as computation and I/O). If the lower 10 address lines go to the same inputs of both chips, then we have to come up with a way to enable

Chip-1 when its address space is accessed (and when $\overline{\text{MEN}} = 0$). Figure 5.28 shows one way we can do this. Notice how the upper 6 lines are wired to the low-active "chip select" ($\overline{\text{CS}}$) input of each chip. We can complete the next step in many ways. Much of it depends on whether more chips are involved. Note how we specify the bundle of digital lines to avoid congesting the schematic diagram: A crossed line with a number next to it represents the number of lines in the bundle.

■ **EXERCISE 5.12**

Wire four TMM2114 (1024 × 4) memory chips using a DMUX.

■ **EXERCISE 5.13**

Wire a 2048 × 8 memory with TMM2114 chips.

5.9.4 Memory Access

Memory is addressed by setting the control lines in a specific order. The sequences to read and write are different, as described next.

Memory Read   Figure 5.29 shows a time sequence of the electronic signals involved in a memory read. For a memory read, the processor follows these steps:

1. The processor sets the address lines to the valid address.
2. The processor sets the read/write line.
3. The processor sets the memory-enable line.
4. When the memory-enable line is asserted, the memory responds within a finite time, called the "access time," by putting the data on the data lines.
5. After a time longer than the access time, the processor automatically reads the content of the data lines, which must contain the data placed by the selected memory chip.

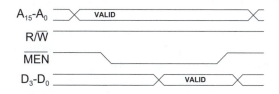

Figure 5.29.   Diagram of logic signals of relevant lines in a memory read. The label "valid" refers to times when the address or data lines have valid information.

Memory Write   For a memory write, the process is similar, except that the data lines are set at the same time that the address lines are set. When the memory enable line is asserted, the selected memory chip reads the contents of the data lines. A diagram of the timing of the electronic signals is illustrated in Figure 5.30. The sequence of steps in the access process is listed next:

1. The processor sets the address lines to the valid address.
2. The processor sets the read/write line.

3. The processor puts the data on the data lines.

4. The processor asserts the memory enable line.

5. The selected memory reads the data lines.

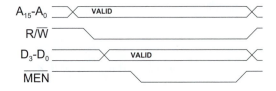

Figure 5.30.   Diagram of logic signals of relevant lines in memory write.

5.10   EPILOGUE TO DIGITAL: DIGITAL I/O

A computer system may control external devices by just setting binary levels on a group of output lines. These may be used for on/off, true/false type of operations, and may involve input as well as output. For example, a set of transducers in an alarm system may just need to specify that the security or safety has been compromised, so a true/false type of digital signal suffices. This signal may also go into a digital I/O output that can be sent back to another analog transducer to turn on the lights, sound the alarm, and if the system is "smart," make decisions and take further steps (such as call the police or the fire department).

The I/O inputs can also be used for counting. For example, some transducers give out a square wave whose frequency contains the information. Such a device could be an optical encoder, with each pulse corresponding to the passage of some optical marker. Thus, some form of digital counting circuit extracts the information, such as position or speed, from the input signal. Other sequential circuits may correlate several inputs, with the timing between them revealing some form of information (for example, more than one stream of encoder outputs may reveal direction of motion in addition to position or speed). Sometimes the inputs may be random except for a fewer number that are correlated. In this case, the circuit may need to correlate the signals of other inputs. For example, the measurement of cosmic rays or elementary particles in physics research involves correlating the inputs from several detectors, which signal the passage of a high-energy particle through them. Correlating the various inputs allows researchers to discriminate real events from detector noise or firings due to other sources. You can also have an alarm system that fires only when certain conditions are met, or a system with a set of data lines that specify more than just the alarm itself, such as the location of the problem and its fail/safe status.

I/O outputs can also be of the on/off type, as in controlling a light indicator or turning a device, such as a door or machine, on or off. Computer-controlled equipment can control the power to an entire city with a simple logic 1 via a digital I/O line (with some redundant system of checks, of course).

Thus, digital I/O lines are binary lines that serve as input or output between the real world and the smart digital system. From the perspective of the computer system, the I/O line may be a bit of a memory location, so it needs to simply read it or store into it for the task to be completed.

The electronic circuits for digital I/O processes belong properly in the analog section because we need to know the internal circuitry and analog specifications of gates. Here we mention two simple cases: input from switches and output to lights. The last chapter covers other I/O circuits that require us to know more about the internal circuitry of gates and ICs.

5.10.1 Application: Digital Input from Switches

In previous sections, we used switches to generate digital inputs. How do they work? We mentioned one example already in this chapter: the SPDT switch. A simpler circuit is that of a push-button switch or a single-pole, single-throw (SPST) switch. Here we want to generate a signal when we temporarily push the button. As mentioned earlier, switches suffer from bouncing, and we must debounce the switch. The previous debounce circuit shown in Section 5.6.3 does not work for a SPST switch. Figure 5.31 shows one way to debounce an SPST switch.

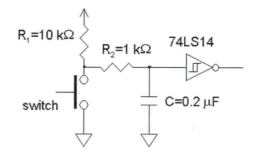

Figure 5.31. Debounced SPST switch using analog RC time constants.

When the switch is not pressed, the input to the NOT gate is a logic 1 via a pull-up resistor connected to 5 V. The output of the gate is a logic 0. When the switch is closed (pushed), the action effectively grounds the input of the NOT gate. We use a 74LS14 Schmitt trigger NOT gate, which has a 1.6 V low to high-threshold and a 0.8 V high to low-voltage threshold. This eliminates noise. When the key is pressed, the output of the NOT gate is a logic 1.

The trick for debouncing is to put the switch in an RC circuit so that the time for charging and discharging the capacitor is longer than the bounce time. In our case, a 100 ms time constant is plenty, but we can lower it to a few milliseconds. In Figure 5.31, the capacitor discharges through R_2 and charges through $R_1 + R_2$. In the circuit shown, the charging time is about 2 ms and the discharge time is 0.2 ms. The resistor R_2 has to be as small when using TTL gates because the current coming out of the gate when the input is low may produce a voltage drop across R_2 that will make the input a logic 1. The capacitor, a high-speed one (such as mica or ceramic), is connected to a Schmitt-trigger inverter to remove false triggerings due to noise in the slow rise and fall of the capacitor voltage.

Figure 5.32 shows another debounce circuit that is more digitally minded and more elaborate, but effective. As before, when the switch is not pressed, the input to the flip-flop is a logic 1 via the pull-up resistor connected to 5 V. This sets the $\overline{Q}$ output of the D flip-flop to 0 when the flip-flop is triggered. The flip-flop is triggered by a square wave with a period of 100 ms. In the example of Figure 5.32, the square wave is generated by an LM555 timer chip (see Section 5.12.1 for more information on this circuit).

When the switch is pushed, the D-input is a logic 0 and the flip-flop output is a logic 1. Any bounces between this input and 100 ms later do not change the state of the flip-flop output because the flip-flop is triggered every 100 ms. The disadvantage of the circuit is

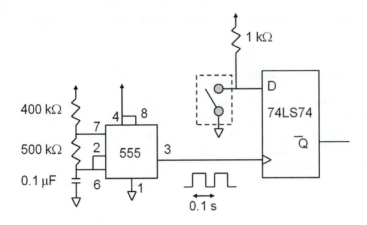

Figure 5.32.   Switch debounce using an oscillator.

that the digital signal produced by the circuit is synchronized with the square wave, which may not be desirable in certain applications, such as games.

5.10.2   Application: Digital Output to Lights

Old science-fiction movies always described computers as complicated devices with lots of lights that blinked actively when the computer was doing something (recall the 1983 movie *War Games*, with a young Matthew Broderick). Each of the lights that came out of the computer, for whatever purpose, was the result of a digital output (in the pretend computer of the movie, Hollywood may have had a different circuit behind them). Not taken to such an extreme, those small lights, light-emitting diodes (LED), are the workhorse of digital outputs (they used to be only red, but now they come in all colors and intensities). LEDs convey a simple and visible yes/no message.

LEDs are diodes that to light properly, need a current of a specific value (about 20 mA) to pass through them in a specific direction. Chapter 9 covers diodes and LEDs, and Chapter 11 covers connecting analog devices with digital signals, so we skip the design details here. For now, it is enough to say that the circuit shown in Figure 5.33 is a good one for driving an LED with a digital gate.

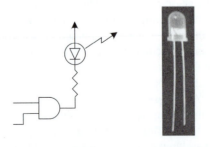

Figure 5.33.   Circuit to drive an LED and a photo of one.

The circuit is simple, but it has some odd features. The LED turns on with a logic 0. This has to do with the way gates are wired: They are good at sinking current but not so good at sourcing current. In series with the LED is a resistor to limit the current (about 180 Ω—again we are getting a little ahead of ourselves, but as with the rest of digital, we can "plug and play" and ask questions later). We must also consider the orientation; in a real LED, the longer lead must be more positive than the shorter lead (if you are unsure, test both ways, and one will work). The circuit is so simple, yet it conveys the purpose: Light on or off is the digital output (1 or 0, true or false). Which one depends on the logic leading to the LED.

Finally, let us give a sense of purpose to our new analog toys. Suppose we want to build a digital game of tic-tac-toe. No computers, just a digital circuit. We need an array of nine regions in a 3 × 3 array. The digital equivalent for × is a red LED, ○ is a green LED. Each region then needs its own circuit, with two switches and two LEDs. Before we select an option (× or ○), both lights must be off. Obviously, we need a flip-flop for each LED. For simplicity we pick D flip-flops. Because a logic 0 activates the light, we connect each to the $\overline{Q}$ output of the corresponding flip-flop. Figure 5.34 shows the circuit for each of the nine regions.

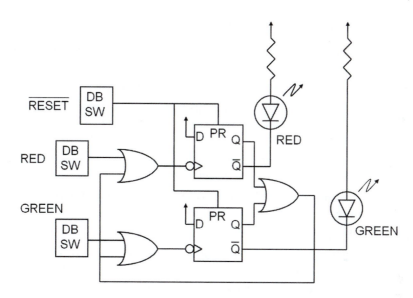

Figure 5.34.   A fun circuit to play a digital version of game tic tac toe.

When the game is over we must turn off all the lights, so we need a global reset button that is connected to the preset input of all the flip-flops. In this design the D-input of a flip-flop is a logic 1, permanently, and the clock input is connected to a (debounced) push-button. This way, when pushing the button, the $\overline{Q}$ output of the corresponding flip-flop becomes a logic 0, turning the LED on. The push button gives out a logic 0 when pushed, so a negative edge-trigger flip-flop is needed. If only positive edge is available then the circuit is activated when we release the button. To prevent the other light from being turned on accidentally, we use a set of OR gates that prevents a flip-flop from being triggered after the other was triggered.

So here we are, ending the digital section with ways to connect to the real physical/visual world via analog. Plug-and-play can go a long way, but then you miss all the fun: analog. So stay tuned and embrace it as our next adventure in electronics. It may get rough, and not so easy, but it will be fun.

5.11 PROBLEMS

1. A second type of RS flip-flop uses NAND gates, as shown in Figure 5.35. This flip-flop differs from the NOR-based flip-flop, in that the inputs are low-active. Write the truth table for the flip-flop.

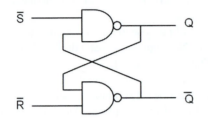

Figure 5.35.   RS flip-flop with NAND gates (Problem 1).

2. Flip flops can be represented by a functional form.

   (a) Show that the output of the RS flip-flop can be represented by the function
   $Q_t = \overline{R} \cdot (Q_{t-1} + S)$.

   (b) Find the functional form of the NAND-based flip-flop of Problem 1.

3. Using positive edge-triggered T flip-flops with clear and preset inputs, design an *asynchronous* decade counter (it counts up to nine and then resets to zero). Make a diagram.

4. Using T flip-flops with clear and preset inputs, design an *asynchronous* decade count-down counter (it counts down to zero and resets to nine after zero). Make a diagram.

5. Show that the circuit of page 37 works as an edge-triggered D flip-flop. Do this the following way.

   (a) Fill in Table 5.15 for $\overline{S}$ and $\overline{R}$.

   (b) Complete the truth table 5.15 for Q and $\overline{Q}$. You may want to do Problem 1 if you have not done so already.

   (c) Make a summary statement describing how the circuit works.

Table 5.15.   Truth Table for Problem 5

D	E	$\overline{S}$	$\overline{R}$	Q	$\overline{Q}$
0	0				
0	1				
0	0				
1	0				
1	1				
1	0				

6. Using a 4-bit binary counter IC, shown in Figure 5.36, design a circuit that outputs a pulse for every six input pulses. Draw a timing diagram of the input clock, counter outputs, and circuit output.

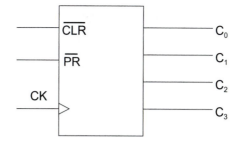

Figure 5.36. Binary counter IC (Problem 6).

7. Design a *synchronous* counter that counts to six and resets to zero after that state. Any unused states should go to zero. Make a diagram. Your design should use JK flip-flops, with J and K not tied together as in a T flip-flop.

8. We are to design a state machine that consists of three negative edge-trigger D-type flip-flops, all synchronously triggered by the same clock pulse. The three outputs of the machine form the digital number $S_2 S_1 S_0$. An incomplete schematic of the circuit is shown in Figure 5.37. When input $x=0$, the circuit makes the outputs change in the following sequence after each clock pulse: $1 \rightarrow 3 \rightarrow 5 \rightarrow 7 \rightarrow 0 \rightarrow 2 \rightarrow 4 \rightarrow 6 \rightarrow 1$. When the external input is $x=1$, the circuit should follow the *reverse* sequence.

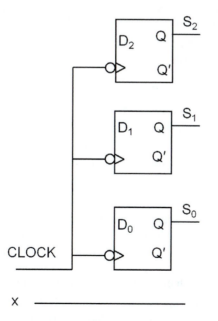

Figure 5.37. Circuit for Problem 8.

(a) Fill in Table 5.16 with the values that the outputs $S_2(t+1)$, $S_1(t+1)$, and $S_0(t+1)$ should have *after* a clock pulse. Two rows of entries have been filled in as an example.

(b) Fill in Table 5.16 with the values that the D inputs of the flip-flops should have so that the outputs change according to the prescribed sequence.

Table 5.16. Table for Problem 8

x	$S_2(t)$	$S_1(t)$	$S_0(t)$	$S_2(t+1)$	$S_1(t+1)$	$S_0(t+1)$	D_2	D_1	D_0
0	0	0	0						
0	0	0	1	0	1	1			
0	0	1	0						
0	0	1	1						
0	1	0	0						
0	1	0	1						
0	1	1	0						
0	1	1	1						
1	0	0	0						
1	0	0	1	1	1	0			
1	0	1	0						
1	0	1	1						
1	1	0	0						
1	1	0	1						
1	1	1	0						
1	1	1	1						

(c) If the flip-flop inputs D_2, D_1, and D_0 are Boolean functions of x, $S_2(t)$, $S_1(t)$, and $S_0(t)$, find the corresponding Karnaugh map and simplified Boolean expressions for each of them.

(d) Complete the circuit of the figure by adding any gates and connections that feed into D_2, D_1, and D_0.

9. Design a programmable four-digit combination lock.

10. Design the circuit of a digital clock.

(a) Design a 1 Hz clock signal using an LM555 timer chip.

(b) Using the 74LS190 decade counter ICs, design the circuitry for driving seven segments for minutes and hours starting from the 1 Hz signal.

(c) Design the circuitry to turn an AM/PM LED.

11. We want to design a safety circuit for a laser lab. The output of the circuit is a shutter line SH that commands an electronic shutter to block a high-power laser. When SH $= 1$ the shutter blocks the laser beam. One logic input to the circuit is the "lab-door" (LD) line. When the door of the lab is open, a sensor in the door frame generates the logic signal LD $= 1$. It is set to LD $= 0$ otherwise. An override (OV) line coming from a push button is 1 while the push button is pressed, 0 otherwise.

(a) Design a logic circuit for the following conditions: (i) When the lab door is closed, the laser beam remains unblocked; (ii) when the door is opened and the push button is not pushed, the laser is blocked; and (iii) when the door is opened and the push button is pressed, the laser is unblocked. Write a truth table and design the circuit.

(b) When the lab door is opened, the laser shutter is activated and the beam is blocked. Modify the circuit so that when the laser beam is blocked, it remains so even if the door is closed or the push button is pressed. Add a new push button input that, when pressed, resets the circuit to its normal operating conditions.

(c) Modify the push-button circuit so that $OV = 1$ for 15 seconds after the push button is pushed, independently of when it is released. (This is so that the laser operator can leave the lab without blocking the laser.)

(d) Design a four-digit combination lock that provides an alternative way of generating $OV = 1$ for 15 seconds. (This keypad will be put outside the lab and will allow the operator to get back into the lab without blocking the laser.)

12. A student wants to redesign the stoplight at the intersection of College Street and Broad Street. It is the only stoplight between campus and off-campus apartments.

(a) Using a 4-bit counter, draw the diagram of a circuit that will turn the lights of the College Street stoplight in the following repetitive sequence: green light for 20 seconds, yellow light for 5 seconds, and red light for 40 seconds. Assume that the input clock frequency is 1 Hz. Assume also that the stoplight will take three high-active inputs, one for each light, that when true, will turn on the corresponding light.

(b) During summer, when College Street has very low traffic, the stoplight switches into the following mode of operation: After finishing the red-light cycle, it will not change to green unless an external input (coming from a car-sensing coil underneath the pavement) is high. Modify the circuit so that it operates in this mode.

(c) A second car-sensing coil is located upstream from the first one so that if the light is green and a car activates it, the green light remains unchanged for at least 5 seconds after the car activates it. This way, the car gets a chance to make it through the intersection.

(d) This semester, you have a late-night activity (academic, of course) that gives you real trouble getting up in the morning. As a consequence, you are always rushing to campus so that you are not late to class. But that College Street stoplight always gets in the way. In a *very* minor transgression from an otherwise fine law-abiding citizen, you decide to add a secret component to the College Street stoplight circuit. It gets activated by remote control so that when you are coming toward the intersection in your car, the light turns to or stays green while you are rushing through the intersection, ensuring the perfect timing for your trip to class. Set your own conditions for this to happen and draw a circuit diagram. *Note*: Your design has to be clear and well explained. Unreadable diagrams and algorithms will be marked wrong even if they work!

13. Design an asynchronous 4-bit binary "down counter" with positive edge-trigger JK flip-flops, with count outputs C_0, C_1, C_2, and C_3.

(a) Draw an accurate timing diagram of the input clock pulse CLK, Q_0, Q_0', Q_1, and Q_1' for six periods of CLK.

(b) Add the necessary wiring to the circuit so that it loads 1100 when it reaches 0000. Assume that the flip-flops have clear and preset (low-active) inputs.

(c) Add the necessary wiring to the circuit such that when an external input START/$\overline{\text{STOP}}$ is 1, the counter counts, and when it is 0, it stops.

(d) Using 2×1 multiplexers modify the wiring so that when an external input $U/\overline{D}$ is 1, the outputs of the circuit C_3, C_2, C_1, and C_0 count up, and when the input is 0, the outputs count down.

14. A calendar clock has a 4-bit month counter. Design a synchronous counter with the following specifications:

    (a) The month count should increase with the rising edge of a TTL pulse that is an input to the circuit.

    (b) The circuit should output a pulse every time the month count goes from month 12 to month 01 (every year, so that it increments a year counter, which is not part of your design).

    (c) The circuit should have four binary outputs that represent the month count.

    (d) You can use only positive edge-trigger JK flip-flops that *do not* have preset or clear inputs.

15. We want to drive a stepper motor in high-torque mode. A stepper motor (see Figure 5.38) has four coils. The motor operates in high-torque mode when its coils are energized consecutively following a four-step cycle that repeats coninuously. Each cycle step corresponds to a motor step. Table 5.17 lists the four-step sequence in which the coil must be energized. A logic 1 denotes that the coil is energized.

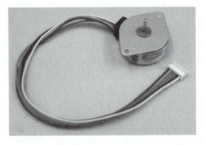

Figure 5.38. View of a stepper motor. It has a series of coils that, when energized, lock the motor shaft in position. Energizing a different set of coils makes the motor step in to a new position.

Table 5.17. Sequence of Energizing Coils to Drive a Stepper Motor (Problem 15)

Step	Coil 1	Coil 2	Coil 3	Coil 4
1	1	0	0	1
2	1	1	0	0
3	0	1	1	0
4	0	0	1	1

The circuit must take an input digital pulse to step a *4-bit synchronous counter* circuit through the sequence specified in the table. The counter must use T-flip flops. Assume that the unused states *never occur*. Each of the counter lines enables a motor coil.

16. Design a circuit that generates the timing cycle of a washing machine.

    (a) Using 4-bit counters, design a circuit that generates a pulse every minute $\left(\frac{1}{60}\text{Hz}\right)$ when the input clock has a 60 Hz frequency. You can use several frequency dividers in tandem if you want.

(b) Suppose that the analog electrical hardware of the washing machine has four low-active inputs: $\overline{FILL}$, $\overline{WASH}$, $\overline{RINSE}$, and $\overline{DRAIN}$. These inputs consecutively activate their respective mechanical functions for the times given in Table 5.18. Design a circuit that has outputs for each of the washing machine cycles. The outputs should be activated consecutively and should last the time given by the table. The input to the circuit is the 1/60 Hz signal of part (a). Figure 5.39 shows the block diagram of the circuit.

Table 5.18.   Timing Cycle of a Washing Machine (Problem 16)

Function	Time (min)
$\overline{FILL}$	2
$\overline{WASH}$	4
$\overline{RINSE}$	4
$\overline{DRAIN}$	2

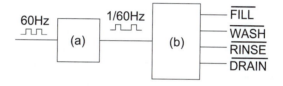

Figure 5.39.   Block diagram for washing machine circuit of Problem 16.

(c) Modify the circuit of the previous part to generate an output every time the washing machine cycle is completed. Use this pulse to reset the washing cycle.

(d) Design a circuit that inhibits the 60 Hz input when the cycle is finished (otherwise, the washing machine will continue washing forever). Include an additional input to restart the cycle (see the block diagram in Figure 5.40).

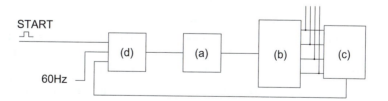

Figure 5.40.   Block diagram of full circuit of Problem 16d.

17. Verify that the 4-bit LFSR goes through all 15 states. Pick any starting value except zero.

18. Using memory-cell blocks like the one shown on the left side of Figure 5.41, construct the internal wiring of a 8 × 2 memory chip shown on the right side of Figure 5.41. Make a diagram.

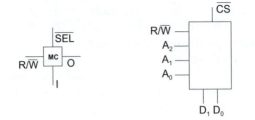

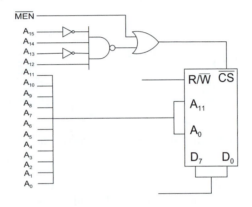

Figure 5.41.   Memory cell (left) to be used to make a 8×2 memory unit (right) in Problem 18.

19. Draw the circuit diagram for a section of memory that contains four 2048×4 memory chips connected to a 16-bit address bus and an 8-bit data bus. Make the proper connections with all the address and data lines so that the starting address is 2000_{16}.

Figure 5.42.   Circuit for Problem 19.

20. For the circuit of Figure 5.42:
    (a) What is the memory capacity of the chip?
    (b) What are the starting and ending addresses?
    (c) By adding another identical memory chip, modify the circuit so that the new circuit has twice the memory capacity as before. Arrange it so that the second chip occupies the memory space following the one of the first chip.

21. The circuit of Figure 5.43 uses an 8-bit comparator chip as part of the address decoding. Determine the memory space of each memory chip in the circuit shown. Explain your reasoning.

22. A memory board addressed by a computer using a 16-bit bus has three memory chips. Use a 4-to-16 line-decoder chip to do the address decoding with the following conditions:
    (a) The first chip has 12 address lines and a starting address $A000_{16}$.
    (b) The second chip has 13 address lines and a starting address $C000_{16}$.
    (c) The third chip's memory space starts at 6800_{16} and ends at $6FFF_{16}$.

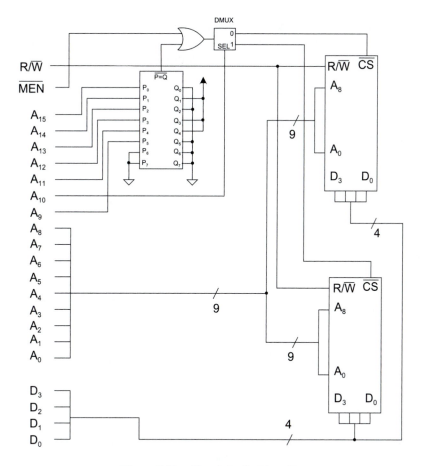

Figure 5.43. Circuit for Problem 21.

5.12   LAB PROJECTS

This section presents ideas for lab activities.

5.12.1   Sequential Circuits

Required ICs: LM555, 74HCT74 (or 74LS74), 74LS76, 74LS190, push-button switch.
Required equipment: Logic board.

The LM555 Timer

Function   The LM555 Timer is a monostable multivibrator IC. It can be configured to operate in several modes, but the most popular are the astable and monostable modes. Here we use the astable mode, and later we go over the monostable mode. In the astable mode, shown next, the LM555 produces a rectangular wave with HIGH and LOW times specified by some external components. Because the inner wirings of the LM555 have some analog components, we will defer covering its description until the chapter on Operational Amplifiers (Chapter 10).

The circuit of Figure 5.44 works the following way: The capacitor C gets charged through R_A and R_B. The LM555 senses V_2, which rises as the capacitor is charged. When $V_2 = \frac{2V_{CC}}{3}$, the LM555 grounds pin 7. This stops the capacitor from charging and provides a path for it to get discharged: Through R_B. When $V_2 = \frac{V_{CC}}{3}$, the LM555 disconnects pin 7, allowing the capacitor to be charged again and repeating the cycle. Pin 3 outputs digital voltages HIGH and LOW while the capacitor is charging and discharging, respectively. It can be calculated (see Section 1.8.5) so that the charging and discharging times are $0.693(R_A + R_B)C$ and $0.693R_BC$, respectively.

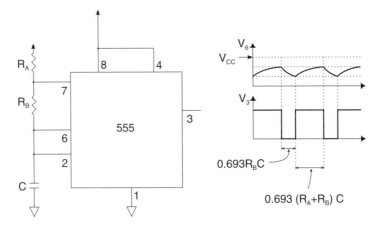

Figure 5.44.   Circuit for lab project 5.12.1.

Pulse Generator   Design and wire a circuit with the LM555 that produces a rectangular wave with a 0.1-second period. The HIGH and LOW times need not be the same.

Frequency Divider   Use the 74LS76, dual, negative edge-trigger JK flip-flop; see the data sheets for pin connections, which for power are not the usual connections (who designed this?). Make a circuit that divides the frequency of the pulse generator by four. Make sure you use 74LS76 and *not* 7476. They are different! Connect the output of the pulse generator to the clock input of the counter. The low-active PRESET and CLEAR inputs of the flip-flops both must be 1 so that they are disabled.

  IMPORTANT: Sequential circuits are sensitive to fluctuations in the power:

- Put a high-speed capacitor (mica or ceramic, $\sim 0.01 - 0.1\mu F$) between power and ground of *every* flip-flop IC. The capacitors stabilize the power to the ICs, preventing accidental triggerings. We go over the reason for this in Chapter 7 (Filters and the Frequency Domain).

- *Never* leave unused *inputs* unconnected. They may cause unwanted transitions.

Counter with the Circuit of the Previous Chapter   Wire a 74LS190 decade counter. Read the data sheets to understand how it works and how to wire it. Modify the circuit with the ALU that you built in a previous lab (see Section 3.11.2): Invert the output S_2 before you connect it to the ALU. This changes the operations that will be produced with the inputs x, y, and z, setting the circuit for $F = B$ with a 000 combination and $F = A$ with a 001 combination. A block diagram of the circuit is shown in Figure 5.45.

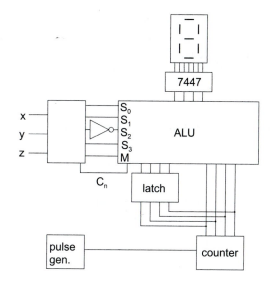

Figure 5.45.   Circuit for lab project 5.12.1

Connect the outputs of the counter to the B inputs of the ALU, and set the x, y and z inputs to the 000 combination. The seven-segment display should now display the value of the counter.

Latching the Counter Value   Now build a 4-bit latch using the 74HCT74 D flip-flops ICs (dual, positive edge-triggered D flip-flop). Use the HCT chip because it is receiving signals from an LS chip. Have an external input (such as a switch from the logic board) be the master latching clock input. Connect the outputs of the counter to the inputs of the latch and connect the outputs of the latch to inputs A of the ALU. This way, as the counter is counting you can "freeze" the value of the counter in the latch at the time the latching clock input was asserted and display that value by setting the ALU to display A.

Modulo-6 Counter   So far, you have an up-counter whose count can be seen through input B of the ALU, and you have a frozen value of it through input A of the ALU. Come up with a way to reset the counter so that it becomes a modulo-6 counter (that is, it goes to zero after 5).

Blinking   If you read the data sheet of the 74LS47 seven-segment driver, you will find that it has an input that disables the display when it is asserted. Use this function to make a blinking display.

Start/Stop   Design a circuit that starts and stops with the transition from low to high of a switch from the logic board.

5.12.2   Memory Access

Lab credit: D. Glenar

Required ICs: TMM2114 memory, DP8216 bus driver.

The objective of this lab experience is to learn to wire a memory chip and its many aspects: addressing and read/write operations. Check the data sheets for the 1024×4 Static RAM TMM2114 chip and for the DP8216 bus driver. Consider some relevant information about them.

Information on the TMM2114 SRAM chip:

- The memory has 10 input address lines (A_0–A_9) that are used to specify the memory location.
- Each memory location has 4 bits. Data lines I/O_0-I/O_3 are used to read or write data to a memory location.
- The input/output character of the data lines is set by the write-enable ($\overline{WE}$) input line, which is low active ($0 =$ write and $1 =$ read).
- The low-active chip-select $\overline{CS}$ line enables the chip.

Information on the DP8216 bus driver:

- It allows the flow of data from the four input DI_0–DI_3 lines to the DB_0–DB_3 lines when the DIEN line is low.
- It allows the flow of data from the DB_0–DB_3 lines to the output DO_0–DO_3 lines when the DIEN line is high.
- The bus driver is enabled with the low-active chip-select line $\overline{CS}$.

Wiring and Testing Memory

1. Wire a circuit with the following requirements:
   - (a) The switches in your logic board specify the address lines A0–A9.
   - (b) Simulate the input data for the memory with switches also.
   - (c) Simulate the output of the memory with LEDs from your logic board.
   - (d) Simulate low-active chip-select ($\overline{CS}$) and read/write ($R\overline{W}$) lines with switches in the logic board.
   - (e) Design a circuit that uses the bus driver to direct the flow of data when reading and writing to memory so that when $\overline{CS} = 0$ and $R\overline{W} = 1$, the data in the memory location specified by A_0–A_9 is displayed on the LEDs, and when $\overline{CS} = 0$ and $R\overline{W} = 0$, the data from the input switches is "stored" in the memory location specified by A_0–A_9.

2. Verify the operation of your memory.

Temporal Messaging The objective of this section is to display consecutive figures on the seven-segment display. The plan is to have a data bit control each segment of a seven-segment display. If you include the period, that makes it eight separate lighting elements of the display. Some details relate to lighting the segments (they turn on with a logic 0) and they need current limiting resistors. In this application, we do *not* use the 74LS47.

Figure 5.46 shows the block diagram of the circuit. We use two TMM2114 memory ICs wired so that we have 8 bits per memory location (you need to figure this out). The memory ICs are buffered by the 6216 buffer. Data inputs go to switches, and data outputs go to segments of the display through resistors. The address lines are connected to the outputs of a 4-bit counter. In a programming stage you can control the memory locations directly

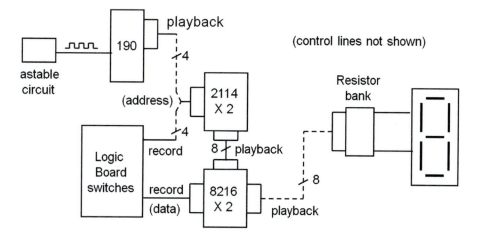

Figure 5.46.   Circuit of lab project 5.12.2.

with switches. In the play stage, the address is controlled by a 4-bit counter driven by an astable signal.

As a first exercise, program the display to show H E L L O in the first five consecutive memory locations the counter steps driven by a 1 Hz signal. As a result, the display will show those letters consecutively. In a second part, try other creative possibilities.

5.12.3   Practicum Test

1. Wire the LM555 IC so that it outputs a sequence of rectangular pulses with a frequency of 0.5 Hz (to better than 50%).

2. Wire the JK flip-flop IC so that it works as a 2-bit up-counter. Its input is the output of the LM555 IC.

3. Wire two LEDs and connect them to the output of the counter so that light-on equals a logic-1.

4. Wire a push button to the clear inputs of the counter flip-flops so that the counter resets when you push the button.

PART II

ANALOG

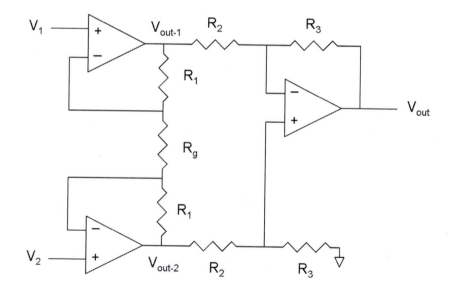

CHAPTER 6

AC SIGNALS

Contents

Analog is at the heart of electronics. It studies the transformation of voltages and currents by devices. This is a huge departure from digital; we will now focus on the multitude of ways in which we can manipulate electrical voltages and currents. Analog electronics well predate digital, and although digital has come a long way in more effectively doing tasks that analog did, we need to get out of that digital cocoon to interface with the real world. Digital works fine to turn devices on or off, but how do we drive those devices and meet their requirements? In digital, we are happy with DC voltages and currents, but that is not how we generate electrical power. Electrical generators at power plants naturally generate alternating voltages and currents. Can we drive devices directly from this source? Yes. Then we need to transport all this electrical power. Are we better off converting to DC?

No. So now we need to take a closer look at how the electronic world works when we use electronics signals that are constantly changing.

Devices respond quite differently to alternating signals, but this gives us new options and tricks. In this analog part of the course we also touch on an important solid-state device that a while ago was a symbol of modern technology: the transistor. This device created a revolution in electronics, changing big electronic boxes with temperamental vacuum tubes into hand-held devices. College-age students do not appreciate this, but the older generation lived a revolution when the radio, a bulky box for a long time, was transformed into a hand-held device: the transistor radio. In the 1960s, this was space-age technology! Not coincidentally, we started building compact flying machines that took us all the way to the moon.

Well, more has happened since then, and we are still living a transition due to (more) modern technology: the computer age, even if it seems it has always been here. The field of electronics proper has also witnessed milestones. We stopped using individual transistors and other components in favor of a more functional electronics box: the operational amplifier. And even in the digital era, technology brought us a new lean and fancy transistor: the MOSFET. This transistor helped speed up the digital revolution, consuming less power and letting us do more things in less space.

So throughout this second part of the course, we visit these iconic devices, and see that they still have a place in electronics today. Our focus is on the popular uses today, avoiding the ones that are currently obsolete. We cover the physics of these devices only slightly, and embarrassingly so for a physicist author, but we have much to do and not enough time to do it.

This chapter gets us back in a steep climb of new concepts and devices. We quickly get into AC circuits, and use one of mathematics' nifty gifts for manipulating signals: complex numbers. We can approach devices that respond to AC signals in many ways, but the "right" way involves using complex numbers. Other treatments skirt the topic with trigonometry or phasors, but that is a lengthy discourse that falls short of treating the matter properly. Embrace the methods and the new algebra. In the end, it will make your understanding much richer and your skills more versatile.

6.1 AC CIRCUITS

This chapter starts by discussing electrical signals that vary with time. The starting point is the simplest type of time-varying signal: the sinusoidal signal. It is also the most important one because any signal, regardless of its shape, can be represented as a superposition of sinusoidal functions of different amplitudes, frequencies, and phases. This chapter is devoted to the response of the simplest components, resistors, capacitors, and inductors, to sinusoidal electrical voltages and currents. Before we start discussing how they respond to time-varying electrical signals, we describe capacitors and inductors in a more formal way.

6.1.1 Representation of AC Signals

Electrical signals that vary sinusoidally with time are represented by the equation

$$V(t) = V_0 \cos(\omega t + \phi) \tag{6.1}$$

where V_0 is the "peak" voltage, ω is the *angular* frequency, and ϕ is the phase of the voltage at $t = 0$. Note that $\omega = 2\pi f$, where f is the frequency normally expressed in Hz (cycles per second). Figure 6.1 shows a graph of this signal as a function of time. The peak voltage in the figure is $V_0 = 4$ V. The horizontal scale underscores that the product of ω and t is phase; it is ωt in units of 2π. The phase offset is 45° or $\pi/4$. Another term that is often used is the "peak-to-peak" voltage, which is the full voltage swing. In the case of a sinusoidal voltage, it is $2V_0$, or 8 V in the example.

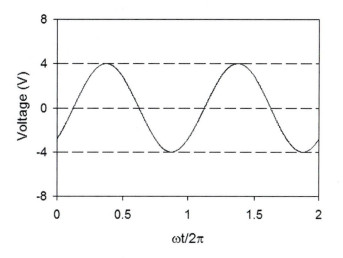

Figure 6.1.   Graph of a sinusoidal voltage.

The root mean square (RMS) value of a time-varying voltage is defined as

$$V_{RMS} = \left[\frac{1}{T} \int_0^T V(t)^2 \, dt \right]^{1/2} \tag{6.2}$$

For a sinusoidal voltage, $V_{RMS} = V_0/\sqrt{2}$. In the example of Figure 6.1, $V_{RMS} = 4/\sqrt{2} = 2.8$ V. A similar relation exists for a time-varying current. The RMS power is

$$P_{av} = \frac{V_{RMS}^2}{R_L} \tag{6.3}$$

which turns out to be the average power delivered to a load resistor R_L. The AC voltage delivered by the standard wall outlet is specified by its RMS value. If the peak-to-peak voltage of a sinusoidal wave is $V_{pp} = 2V_0$, then 110 V AC in our wall outlets represent sinusoidal voltages with peak-to-peak swings of 311 V!

When we have signals that vary by orders of magnitude, it is more convenient to use a logarithmic unit: the decibel, or dB. It is always defined in terms of the ratio of two magnitudes. For voltages, it is defined as

$$\left| \frac{V_1}{V_2} \right|_{dB} = 20 \log \frac{V_1}{V_2} \tag{6.4}$$

For power, it is defined as

$$\left| \frac{P_1}{P_2} \right|_{dB} = 10 \log \frac{P_1}{P_2} \tag{6.5}$$

This notation is used a lot to express attenuation or gain of a signal being processed, where $V_{out} = V_1$ and $V_{in} = V_2$. Table 6.1 gives a few handy values. Because factors of $\sqrt{2}$ and 2 are common in many situations, remember the corresponding decibel values: 3 dB and 6 dB, respectively.

Table 6.1.   Popular Decibel Conversions

V_1/V_2	In dB
0.01	−40
0.1	−20
0.5	−6
0.707	−3
1	0
1.414	+3
2	+6
4	+12
10	+20
100	+40

Before we continue with sinusoidal voltages, let us consider another type of time-varying voltage: the rectangular wave. A particular type of rectangular voltage signal switches between two values, a high voltage of V_0 and low voltage of 0 V. The period may be divided into uneven parts of duration t_{hi} and t_{lo}, as shown in Figure 6.2. The duty cycle is a useful parameter for this type of signal. It is defined as

$$D = \frac{t_{hi}}{(t_{hi} + t_{lo})} \tag{6.6}$$

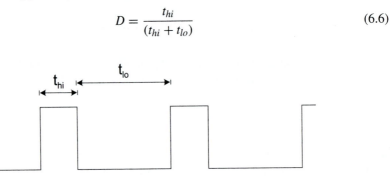

Figure 6.2.   A rectangular waveform.

It can easily be shown that the average voltage is

$$V_{av} = V_0 D \tag{6.7}$$

and the RMS voltage is

$$V_{RMS} = V_0 \sqrt{D} \tag{6.8}$$

The average power delivered to a load R_L is given by Equation 6.3. For a rectangular wave, it is

$$P_{av} = \frac{V_0^2 D}{R_L} \tag{6.9}$$

6.1.2 Capacitor in an AC Circuit

We have seen that the current in a capacitor C is related to the voltage across it via Equation 1.27. If we apply a voltage

$$V(t) = V_0 \sin \omega t \tag{6.10}$$

to the capacitor in the circuit of Figure 6.3, then we can obtain the current flowing through it:

$$I(t) = \frac{dq}{dt} \tag{6.11}$$

$$= C\frac{dV}{dt} \tag{6.12}$$

$$= V_0 \omega C \cos \omega t \tag{6.13}$$

$$= V_0 \omega C \sin(\omega t + \pi/2) \tag{6.14}$$

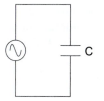

Figure 6.3. Capacitor driven by a sinusoidal voltage.

We note that the capacitor is an open circuit, but as charge flows into one plate and out of the other plate, we can say that the current is effectively flowing through it. Even in the first chapter, where we referred to the charging of the capacitor, current flows into the capacitor to charge one plate and out of it to charge the other plate with the opposite charge, so the statement is not really incorrect. Notice Figure 6.4. As shown in Equation 6.14, the current and the voltage are 90° *out of phase*. This aspect of AC circuits is the hardest

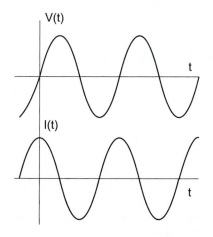

Figure 6.4. Voltage and current for a capacitor driven by a sinusoidal voltage, such as the circuit of Figure 6.3.

one to understand. Because the current is the derivative of the voltage, then the voltage "lags" the current, as shown in the figure. We can understand this better by remembering the charging of the capacitor seen in the first chapter: When we start charging the capacitor (that is, when the voltage across it is zero), the current is maximum; when the voltage is maximum (that is, when the capacitor is fully charged), the current is zero. That applies to sinusoidal signals as well.

We also see an interesting result: The amplitude of the current is $I_0 = V_0\omega C$. Notice that it depends on the *frequency*. If we define a resistance or "capacitive reactance" for the capacitor á la Ohm's law, then it is $R_C = 1/\omega C$. This frequency dependence has interesting consequences: When the frequency is high, its reactance is very small. Another way to view it is that because the current is the rate of change of the voltage across the capacitor, then the faster the voltage varies, the higher the current, and hence a low reactance. At the other end of the frequency spectrum, when the frequency is zero, the capacitive reactance is infinite. That is, when the voltage does not change, the current is zero. This makes sense because a capacitor is an open circuit for DC signals.

We saw earlier in the digital part of the course that we can eliminate the effect of voltage spikes on the voltage supply of a flip-flop IC by putting a (shunt) capacitor between the power and ground pins of the IC. The spikes are a high-frequency signal, so they get an easier path to ground through the capacitor than through the IC. As you will see later, this frequency-dependent resistance is the basis of filters.

6.1.3   Inductor in an AC Circuit

Equation 1.15 relates the current flowing through an inductor to the voltage across it. Consider the circuit of Figure 6.5. If the current flowing through the inductor is given by

$$I(t) = I_0 \sin(\omega t - \pi/2) \tag{6.15}$$

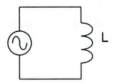

Figure 6.5.   Inductor driven by a sinusoidal voltage.

then the voltage is

$$V(t) = L\frac{dI}{dt} \tag{6.16}$$
$$= I_0\omega L \cos(\omega t - \pi/2) \tag{6.17}$$
$$= I_0\omega L \sin(\omega t) \tag{6.18}$$

Here we see also that the voltage and currents are out of phase by $\pi/2$, but with the phase shifted in the opposite direction compared to the case for the capacitor. This is shown in Figure 6.6, where we can see that now the voltage "leads" the current. Again, this is because one is the rate-of-change of the other.

Similarly to the case of the capacitor, we can define an inductive reactance for the inductor: $R_L = \omega L$. Notice that it also depends on the frequency, but now it is directly

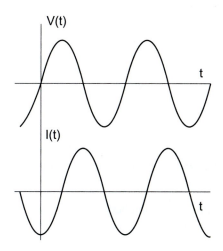

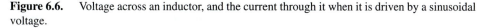

Figure 6.6.   Voltage across an inductor, and the current through it when it is driven by a sinusoidal voltage.

proportional, the opposite of the case for the capacitor. At high frequencies, inductors have high reactances, and at zero frequency they have zero reactance (acting like a straight wire). Because of the high reactance at high frequencies, they are useful as surge protectors; when in series with a power line, they suppress high-frequency spikes.

6.1.4 Complex Numbers

As we have seen, capacitors and inductors have reactances that depend on the frequency. We have also seen the existence of a phase difference between the current and voltage, but the reactances do not account for it. Complex numbers help in this regard by unifying the phase and amplitude, thus providing a complete description of components in AC circuits.

Before we get into discussing AC circuits with complex numbers, it is useful to do a crash refresher of complex numbers. This is not a thorough description, but it brings the necessary elements of the formalism for understanding the topics that follow.

One way to look at complex numbers is to view them as vectors on a plane, shown in Figure 6.7. A complex number is defined in general as

$$z = a + jb \tag{6.19}$$

where a is the "real" part and b is the "imaginary" part. The imaginary number "j" is defined as $j = \sqrt{-1}$ (devious but clever, as you will see later).[1] In our vector analogy, a is the x-component and b is the y-component of the vector. The modulus is

$$|z| = \sqrt{a^2 + b^2} \tag{6.20}$$

and the angle that the vector forms with the x-axis is

$$\theta = \tan^{-1}(b/a) \tag{6.21}$$

---

[1]In most mathematical treatments, the imaginary number is represented by i but in electronics, it is the symbol for the current. Therefore, throughout electrical engineering the symbol j is used instead.

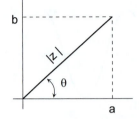

Figure 6.7. Vector in the complex plane.

We can do operations with complex numbers just as we would with vectors. For example, suppose we have two complex numbers, $z_1 = a + jb$ and $z_2 = c + jd$. Then

$$z_1 + z_2 = (a + c) + j(b + d) \tag{6.22}$$
$$z_1 - z_2 = (a - c) + j(b - d) \tag{6.23}$$

However, you can also multiply complex numbers. You can verify that

$$z_1 z_2 = (ac - bd) + j(ad + bc) \tag{6.24}$$

In arriving at the previous expression, we used $j \cdot j = -1$. It is also useful to define the complex conjugate of z as

$$z^* = a - jb \tag{6.25}$$

■ **EXERCISE 6.1**

Verify the following relationships:

1. $zz^* = a^2 + b^2$
2. $z + z^* = 2a$
3. $z - z^* = j2b$
4. $(z^*)^* = z$

We can also define division of complex numbers:

$$\frac{z_1}{z_2} = \frac{(ac + bd) + j(bc - ad)}{c^2 + d^2} \tag{6.26}$$

where we have used

$$\frac{z_1}{z_2} = \frac{z_1 z_2^*}{z_2 z_2^*} \tag{6.27}$$

to keep the denominator real (see Exercise 6.1).

An alternate way to represent a complex number z is

$$z = |z|(\cos\theta + j\sin\theta) \tag{6.28}$$

Using the Euler's identity

$$e^{j\theta} = \cos\theta + j\sin\theta \tag{6.29}$$

we can express z as

$$z = |z|e^{j\theta} \tag{6.30}$$

Using the latter notation, complex number multiplication and division is straightforward:

$$z_1 z_2 = |z_1||z_2|e^{j(\theta_1+\theta_2)} \tag{6.31}$$

and

$$\frac{z_1}{z_2} = \frac{|z_1|}{|z_2|}e^{j(\theta_1-\theta_2)} \tag{6.32}$$

where θ_1 and θ_2 are obtained using the definition of Equation 6.21. It can also be shown that $z^* = |z|\exp(-j\theta)$.

For example, if $z_1 = 1 + 2j$ and $z_2 = 1 + j$, then $|z_1| = \sqrt{5}$, $|z_2| = \sqrt{2}$, $\theta_1 = 0.35\pi$, and $\theta_2 = 0.25\pi$. Then $z_1/z_2 = \sqrt{5/2}\,\exp(0.32j)$, where the phase is always expressed in radians.

6.1.5 Redefinition of Reactances

In general, a circuit that has inductors, resistors, and capacitors has voltages and currents with different amplitudes and phases. It is therefore a big challenge to keep track of all those amplitudes and phases. One method of solving AC circuits is to apply the definitions of Equations 6.12 and 6.16. However this method is useful only for very simple circuits. A better method involves the use of complex numbers. Its shortcoming is the abstraction of complex numbers, but in exchange we get a much more straightforward solution. If we define the resistances and reactances, more generally called *impedances*, to be complex, we can get all the amplitude and phase information by solving circuits that treat capacitors and inductors (and resistors) as impedances.

Normally, the input voltage of a circuit has the form $V(t) = V_0\cos(\omega t)$. This is the real part of

$$V = V_0 e^{j\omega t} \tag{6.33}$$

One of the great tricks of this method is that we can do all the algebra assuming that the input voltage is given by Equation 6.33 and then when we reach the answer, we can convert back to the "real" parts (which is what we actually measure). Also note that if we want to graph voltages or currents, we need to get the real part of their complex expression.[2]

Let us start our complex treatment of electrical components with the capacitor. If we replace Equation 6.33 into $I = C\,dV/dt$ (Equation 6.12), we get the current to be

$$I = j\omega C V_0 e^{j\omega t} \tag{6.34}$$

Using Equation 6.29, you can also verify that

$$j = e^{j\pi/2} \tag{6.35}$$

Replacing Equation 6.35 into Equation 6.34 we get

$$I = \omega C V_0 e^{j(\omega t+\pi/2)} \tag{6.36}$$

---

[2]A common beginner's error is to try to plot Equation 6.33. Because the function is a complex number, the graphing software will not know what to do. You must always graph the real or imaginary components of a complex function.

which makes it consistent with Equation 6.14. Similarly to Equation 6.33, the argument of the exponential specifies the phase of the current. It accounts for the $\pi/2$ phase difference between the voltage and current that we mentioned before. If we allow the impedance of the capacitor to be complex, then by comparing Equation 6.34 to Ohm's law, we get

$$Z_C = \frac{1}{j\omega C} = -\frac{j}{\omega C} \tag{6.37}$$

We can obtain the complex impedance of the inductor in a similar way. It is

$$Z_L = j\omega L \tag{6.38}$$

A resistor has a current that is in phase with the voltage across it, so the impedance of a resistor is

$$Z_R = R \tag{6.39}$$

■ **EXERCISE 6.2**

Show that the complex impedance of the inductor is given by Equation 6.38. Assume that the current through the inductor is $I = I_0 \exp[j\omega t]$.

6.1.6   Generalized Ohm's Law

If we generalize Ohm's law for complex impedances, we get

$$I = \frac{V}{Z} \tag{6.40}$$

If the impedance Z is given by the general form

$$Z = |Z|e^{j\theta} \tag{6.41}$$

then by putting the voltage in the form of Equation 6.33, we get the current to be

$$I = \frac{V_0}{|Z|}e^{j(\omega t - \theta)} \tag{6.42}$$

This is a complex number with an amplitude $V_0/|Z|$ and a phase $\omega t - \theta$. The real part is given by $I(t) = (V_0/|Z|) \cos(\omega t - \theta)$.

We can summarize the method in the following way: Convert the voltage to the exponential notation, do all the algebra in that notation, and then get the current and other voltages by taking the real part of the answers.

When we have impedances in series or in parallel, we can get the equivalent impedance the same way as for resistors. When they are in series (Figure 6.8), the sum of impedances is the equivalent impedance:

$$z_{eq} = z_1 + z_2 \tag{6.43}$$

When they are in parallel (Figure 6.9), the sum of the reciprocals is the reciprocal of the equivalent impedance:

$$\frac{1}{z_{eq}} = \frac{1}{z_1} + \frac{1}{z_2} \tag{6.44}$$

Note that because the impedance of capacitors is inversely proportional to the capacitance,

Figure 6.8.   Impedances in series.

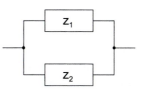

Figure 6.9.   Impedances in parallel.

the formulas for series and parallel combinations are the opposite of resistors, as seen earlier in the first chapter.

We can also generalize our voltage divider to complex impedances, as shown in Figure 6.10. The output voltage is given by

$$V_{out} = \frac{z_2}{z_1 + z_2} V_{in} \tag{6.45}$$

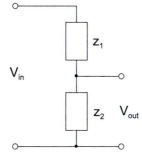

Figure 6.10.   Voltage divider for impedances.

6.1.7   Dissipated Power

The average power dissipated in a circuit is given by

$$P_{av} = \frac{1}{T} \int_0^T I(t)V(t)dt \tag{6.46}$$

This equation reduces to

$$P_{av} = \frac{1}{2} Re(V^* I) \tag{6.47}$$

If the circuit in question has an impedance

$$Z = |Z| e^{j\theta} \tag{6.48}$$

then the phase difference between I and V is θ. We leave it as an exercise for the reader to show that the average power is

$$P_{av} = \frac{V_0^2}{2|Z|} \cos \theta \qquad (6.49)$$

The term $\cos \theta$ is called the power factor, which specifies how much of the delivered power is actually dissipated. From Equation 6.49, you can see that for a resistor, all the delivered power is dissipated (because $\theta = 0$). Conversely, a capacitor dissipates no power because $\theta = \pi/2$. One way to look at it is to think about power being delivered. Is the power supply transferring any energy in the form of heat, light, motion, audio, and so on? The less it does, the more we are not using the power of the supply. For example, a motor has many coils that make it mostly inductive, but the energy delivered to the loads that it moves appears as a resistive component. When $\theta = \pi/2$, a purely capacitive or inductive circuit, we are not delivering power; when $\theta = 0$, a purely resistive circuit, we are delivering the most power.

6.1.8   Worked Example

As an example, we solve the LR circuit shown in Figure 6.11. Assuming that the voltage source is $V_0 e^{j\omega t}$, we want to know (1) the current flowing through the circuit and (2) the voltage across the resistor.

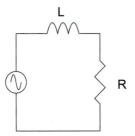

Figure 6.11.   RL circuit driven by a sinusoidal voltage.

Current   Applying Ohm's law, we get

$$I = \frac{V_0 e^{j\omega t}}{Z_T} \qquad (6.50)$$

where

$$Z_T = Z_R + Z_L \qquad (6.51)$$

Replacing the corresponding values of the impedances, we get

$$Z_T = R + j\omega L = \sqrt{R^2 + \omega^2 L^2} \, e^{j\theta} \qquad (6.52)$$

where

$$\theta = \tan^{-1}\left(\frac{\omega L}{R}\right) \qquad (6.53)$$

Therefore,

$$I = \frac{V_0}{\sqrt{R^2 + \omega^2 L^2}} \, e^{j(\omega t - \theta)} \qquad (6.54)$$

Voltage Across R   Applying Ohm's law to R, we get

$$V_R = RI = \frac{V_0 R}{\sqrt{R^2 + \omega^2 L^2}} e^{j(\omega t - \theta)} \tag{6.55}$$

The ratio of the amplitudes of the voltage across the resistor to the input voltage is

$$\frac{|V_R|}{|V_0|} = \frac{R}{\sqrt{R^2 + \omega^2 L^2}} \tag{6.56}$$

Figure 6.12 graphs this result as a function of $\omega/(L/R)$.

When $\omega \to 0$, we have $|V_R|/|V_0| \to 1$, and when $\omega \to \infty$, we have $|V_R|/|V_0| \to 0$. If we consider the voltage across the resistor as the output, then because the circuit attenuates the input at high frequencies, it acts as a low-pass filter, as shown in the graph of Figure 6.12.

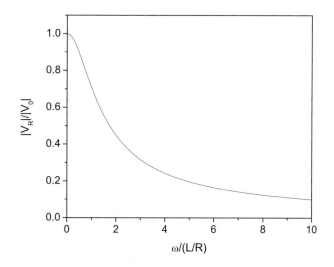

Figure 6.12.   Ratio of the magnitudes of output and input voltages as a function of signal frequency.

Alternatively, if we wanted to calculate only the voltage across the resistor, we would not need to find the current. We could use the voltage divider equation

$$V_R = \frac{Z_R}{Z_R + Z_L} V_0 e^{j\omega t} \tag{6.57}$$

or

$$V_R = \frac{R}{R + j\omega L} V_0 e^{j\omega t} \tag{6.58}$$

Using the exponential notation the last equation becomes

$$V_R = \frac{R}{\sqrt{R^2 + \omega^2 L^2}} V_0 e^{j(\omega t - \theta)} \tag{6.59}$$

where θ is given by Equation 6.53. From this we also see that the voltage across the resistor is out of phase from the input voltage by an amount θ. It is also important to note that θ is frequency dependent, tending to zero when $\omega \to 0$ and to $\pi/2$ when $\omega \to \infty$.

6.2 EQUIVALENT CIRCUITS

A couple fundamental theorems follow closely behind Kirchoff's laws and voltage dividers in importance. They give a way to condense an entire circuit into an equivalent one with only two components: a source and an impedance. The first one, Thevenin, puts the equivalent circuit in terms of a voltage source, and the second one, Norton, does it in terms of a current source.

6.2.1 Thevenin's Theorem

Thevenin's theorem states that *a two terminal network of impedances and voltage sources is equivalent to a single impedance in series with a voltage source.* Figure 6.13 shows this schematically. The Thevenin parameters are obtained the following way:

1. The Thevenin voltage V_{Thev} is the open circuit voltage.

2. The Thevenin impedance can be calculated by two methods:

    (a) Z_{Thev} is the equivalent impedance between points A and B after all the voltage sources have been replaced by straight wires and all current sources replaced by open circuits; or

    (b)

$$Z_{Thev} = \frac{V_{Thev}}{I_{short}} \tag{6.60}$$

   where I_{short} is the current flowing through A and B when we connect those two points with a straight wire.

Method (a) is more convenient because method (b) involves solving the whole circuit. However, method (b) also lets us calculate the Norton equivalent circuit, as we see next.

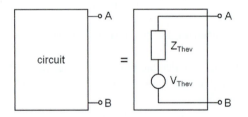

Figure 6.13. Thevenin equivalence synthesis.

6.2.2 Norton's Theorem

In some circumstances, using this theorem is more convenient because it puts the equivalent circuit in terms of a current source. The Norton equivalent circuit (see Figure 6.14) is a current source with an impedance in parallel:

1. The Norton impedance is

$$Z_{Norton} = Z_{Thev} \tag{6.61}$$

It is found the same way as the Thevenin impedance.

2. The Norton equivalent current is

$$I_{Norton} = I_{short} = \frac{V_{Thev}}{Z_{Norton}} \qquad (6.62)$$

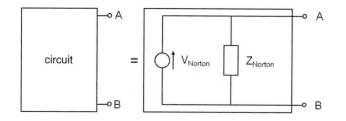

Figure 6.14. Norton equivalence synthesis.

Worked Example These theorems sound pretty dense, so we will do an example. Consider the circuit of Figure 6.15. For simplicity, we use DC supplies and resistors, but the theorem can be generalized to complex impedances. The equivalent resistance between the output points when the source is shorted reduces to the three resistances in parallel, so $R_{Thev} = 1\,\text{k}\Omega$. The resistance of the two vertical branches simplifies to $1.5\,\text{k}\Omega$ (they are in parallel). For the Thevenin voltage, we apply the voltage divider equation and get $(15\,\text{V})(1.5\,\text{k}\Omega)/(1.5\,\text{k}\Omega + 3\,\text{k}\Omega) = 5\,\text{V}$. The Thevenin equivalent circuit is then a 5-V source in series with a resistance of $1\,\text{k}\Omega$. For the Norton equivalent, we connect A and B with a straight wire and find the current to be $I_{Norton} = (15\,\text{V})/(3\,\text{k}\Omega) = 5\,\text{mA}$. Alternatively, $I_{Norton} = V_{Thev}/R_{Thev} = (5\,\text{V})/(1\,\text{k}\Omega) = 5\,\text{mA}$. The Norton equivalent circuit then is a current source of 5 mA in parallel with a resistance of $1\,\text{k}\Omega$.

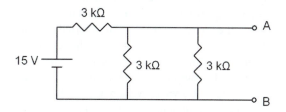

Figure 6.15. Circuit for worked example.

6.3 CIRCUIT LOADING

An important consideration in the design of circuits is the load to which it is connected. In this section we discuss two types of loading. They differ in subtle ways. When you design a circuit you should ask yourself what aspect of the output of the circuit you want to preserve once a load is attached to it. This can be either the signal (a voltage), or the overall electrical power.

6.3.1   Maximizing Signal Transfer

Any circuit can be represented by a voltage source or voltage signal (V_{Thev}) in series with an impedance (Z_{Thev}). If we connect the output of such a circuit to a load with impedance Z_L, as shown in Figure 6.16, then the voltage applied to the load (V_{AB}) is always less than V_{Thev}. If we think about it, this circuit is a voltage divider, with the divided voltage V_{AB} given by

$$V_{AB} = \frac{Z_L}{Z_{Thev} + Z_L} V_{Thev} \tag{6.63}$$

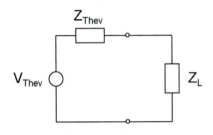

Figure 6.16.   A synthesis of any circuit connected to a load.

Therefore, if we want to get as much of the voltage source or signal as possible (that is, $V_{AB} \sim V_{Thev}$), then

$$R_L \gg Z_{Thev} \tag{6.64}$$

The latter is an extremely important result to keep in mind when designing circuits. A useful rule of thumb is to aim for:

$$Z_L \geq 10 Z_{Thev} \tag{6.65}$$

All power supplies and batteries behave as if they have an internal resistance, regardless of whether they actually have one. That is, the voltage of the source without a load connected to it is different from the voltage when a load has been connected. The difference can be accounted for by modeling the source as if it had an internal resistance. For example, if the source is modeled as an ideal source of voltage V_s in series with an internal source resistance R_s (assuming the purely resistive case), then when the source is unconnected it outputs V_s, but when connected, it outputs $V_s - IR_s$. The latter is reduced to the Thevenin voltage of the source $V_s R_L/(R_s + R_L)$ when we replace $I = V_s/(R_s + R_L)$. In most cases, the internal resistance is small. For example, ordinary batteries have internal resistances of about a fraction of an ohm. This resistance increases with the use of the battery, and also depends on temperature. You can find more information on this in the manufacturer's specifications. All supplies are constructed with a low internal impedance in mind.

6.3.2   Maximizing Power Transfer

The previous arguments and criteria refer to maximizing the transfer of an electrical signal. However, if we want to maximize the power transfer, the criterion changes. Sticking to the purely resistive case, the power delivered by a source is

$$P_s = \frac{V_s^2}{(R_s + R_L)} \tag{6.66}$$

However, the power dissipated by the load can be calculated as

$$P_s = \frac{V_s^2 R_L}{(R_s + R_L)^2} \qquad (6.67)$$

Figure 6.17 shows a graph of the power dissipated by the load as a function of the load resistance R_L, for fixed V_s and R_s.

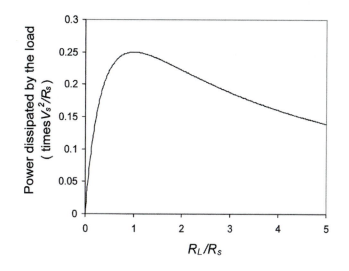

Figure 6.17. Power dissipated by the load as a function of the load resistance.

Notice that dissipated power goes through a maximum at $R_L = R_s$. In this case, the power dissipated by the load is $P_L = V_s^2/4R_s$, which is half the power delivered by the source $P_s = V_s^2/2R_s$. This situation is required in many applications. For example, when signals are broadcast via antennas, we want to maximize the transmitted power, so the impedance of the antenna (generalizing now to complex impedances) must be matched to the impedance of the source. An important part of the design is then the impedance-matching apparatus. When the loads are complex, we need to match the complex-conjugate of the source impedance.

Finally, in high-frequency applications, signals transmit through cables like waves. If the impedances of the transmission line and the load do not match, then the signal wave gets reflected. Thus, in these circuits we must match the source, the transmission line, and the load. In most signal applications the impedance of choice is 50 Ω. Function generators normally have an output impedance of 50 Ω, and standard coaxial cables (RG58) have an impedance per unit length of also 50 Ω. To avoid reflections at the end of the line (cable), the impedance of the end (a piece of equipment or another circuit) must be matched to the impedance of the line. So the input impedance of the end device must also be 50 Ω.

This is a common occurrence with pulsed circuits. We saw in the digital section that we can easily be sending nanosecond-wide pulses in a given application. The rule of thumb is that if the duration of the pulse is comparable to the time it takes the signal to propagate from end to end of the cable (about 1.5 ns/ft), then we must worry about reflections. We can easily see this with a pulse generator feeding to a coaxial cable and ending on an oscilloscope. If the input impedance of the oscilloscope is the standard 1 MΩ, then we will see distorted pulses, or ringing (oscillations). By setting the input impedance of the scope to 50 Ω, or if not possible, attaching a 50 Ω terminator connected to the oscilloscope input with a coaxial tee, we can recover the original shape of the signal (for an example, see Lab Project 6.5.4). Other networks have other impedances as well. For example, analog video and cable-TV signals have a 75-Ω impedance, and radio antenna split-wire cables are 300-Ω networks.

6.4 PROBLEMS

1. An electrical signal oscillates between 0 and V. Find the RMS value of the wave if the wave is

    (a) Square
    (b) Sinusoidal
    (c) Triangular

2. If $z_1 = 1 + j$, $z_2 = 1 - j$, and $z_3 = -2j$, calculate the magnitude and phase (in radians) of

    (a) z_1
    (b) z_2
    (c) z_3
    (d) $z_1 + z_2$
    (e) $z_1 - z_3$
    (f) $z_1 z_2$
    (g) $\frac{z_1 + z_2}{z_2}$
    (h) $\frac{z_1}{z_3}$
    (i) $\frac{z_2}{z_1 - z_3}$
    (j) $\frac{z_3}{z_1} + z_2$

3. If the angular frequency of the source is 10^5 rad/s, calculate the equivalent impedance between A and B in Figure 6.18.

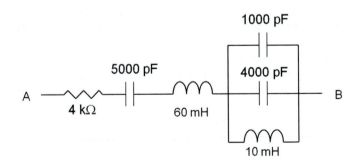

Figure 6.18.   Circuit for Problem 3.

4. Consider the circuit of Figure 6.19. The AC supply has a peak voltage of 10 V and a frequency of 1 kHz. The resistor has a value of 1 kΩ, and the capacitor has a capacitance of 0.1 μF.

    (a) Find the peak value of the current through the circuit.
    (b) Explain how the current flows through the capacitor.
    (c) What is the peak voltage between points A and B?

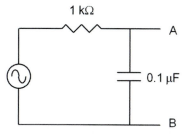

Figure 6.19.   Circuit for Problem 4.

(d) We change the frequency of the source so that the peak voltage through the capacitor is the same as the peak voltage through the resistor

   i. What is the value of the frequency?

   ii. What are the peak voltages across the resistor and capacitor?

   iii. Explain why they are not 5 V.

5. Consider the circuit of Figure 6.20. A capacitor with $C = 0.25\,\mu\text{F}$ is in series with two resistors, $R_1 = 1500\,\Omega$ and $R_2 = 3500\,\Omega$. The peak current through the circuit is $I_0 = 2\,\text{mA}$, and its angular frequency is $\omega = 180\,\text{rad/s}$.

(a) Find the voltages on each component (amplitude and phase).

(b) Find the amplitude of the voltage of the source.

(c) Find the impedance of the circuit.

(d) Find the phase between the source voltage and the current.

(e) Write the expressions for the voltage and current as a function of time (amplitudes and phases all included explicitly).

(f) What is the average power dissipated by the resistors?

(g) What is the average power dissipated by the source?

(h) Explain the relationship between the results in (f) and (g).

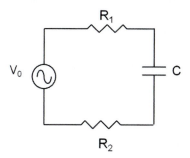

Figure 6.20.   Circuit for Problem 5.

6. In the circuit of Figure 6.21, the voltage source is $V_G = 10\sin\omega t$ V, where the frequency of the source is 20 kHz.

(a) Find the current through the resistor.

(b) Find the current through the inductor.

(c) Find the amplitude of the total current flowing out of the source.

(d) Find the phase between the voltage of the source and the current flowing through it.

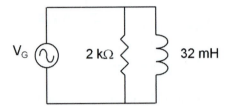

Figure 6.21.   Circuit for Problem 6.

7. The input voltage for the circuit shown in Figure 6.22 is given by $V_G(t) = V_0 \cos \omega t$.

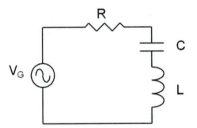

Figure 6.22.   Circuit for Problem 7.

(a) Find the equivalent impedance of all three components in the complex notation.

(b) What is the expression for amplitude of the current?

(c) What is the expression for the phase difference between the current and the voltage of the source?

(d) If $V_0 = 10$ V, $R = 10$ kΩ, $L = 2.5$ H, and $C = 0.01$ µF, make a plot (using Excel or your favorite software) of the amplitude of the current vs. frequency (f not ω).

(e) Make a plot of the amplitude of the voltage across the resistor as a function of frequency.

(f) If we assume V is an input voltage and V_R is the output voltage, what utility can you find for this circuit?

8. In the circuit of Figure 6.23:

(a) If the amplitude of the voltage is 20 V, calculate the current flowing through each element when the angular frequency is 100 rad/s.

(b) Calculate the frequency at which the current leaving the supply is maximum.

(c) Calculate the phase between the current and voltage for parts 8a and 8b.

9. In the circuit of Figure 6.19, the supply has a peak voltage of 10 V and a frequency of 1 kHz. The resistor has a value of 1 kΩ, and the capacitor has a capacitance of 0.1 µF.

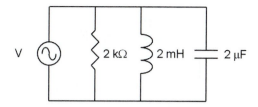

Figure 6.23. Circuit for Problem 8.

(a) Find the Thevenin equivalent circuit.

(b) If we varied the frequency of the source, at what frequency would the output impedance be:

    i. Maximum

    ii. Minimum

    iii. What are the impedances for cases 9(b)i and 9(b)ii?

(c) We now connect a resistor R in series with a 0.02 μF as a load connected to points A and B.

    i. What value of resistor R will not load the circuit when the frequency is 1 kHz?

    ii. What value of resistor R will not load the circuit regardless of frequency?

10. Suppose we want to apply a voltage of 2 V ($\pm$ 10%) to a 10 kΩ resistor, but we have only a 10 V power supply. Design a voltage divider that will give 2 V from the 10 V power supply. Choose the resistors of the divider so that: (a) the maximum current put out by the supply does not exceed 100 mA, and (b) when the divider is connected to the 10 kΩ resistor, the voltage does not drop below 1.8 V (a 10% loading of the divider).

11. Find the Thevenin equivalent circuit between points A and B for the circuit shown in Figure 6.24.

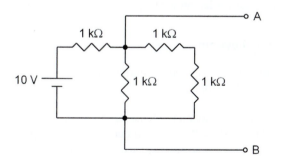

Figure 6.24. Circuit for Problem 11.

12. For the circuit of Figure 6.25:

(a) Find the Thevenin equivalent circuit between A and B.

(b) Determine what value of load resistance applied to A and B will not load the circuit.

(c) Find the Norton equivalent circuit.

(d) What load value *maximizes* the delivered power?

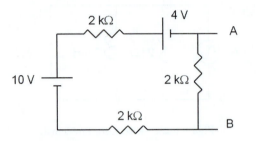

Figure 6.25. Circuit for Problem 12.

13. Use the Thevenin equivalent of the circuit of Figure 6.26 to find the voltage that is applied to a 3 kΩ load connected across A and B.

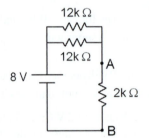

Figure 6.26. Circuit for Problem 13.

6.5 LAB PROJECTS

This section presents ideas for lab projects.

6.5.1 Circuits and Thevenin

Required equipment: DC power supply, multimeter.
Required components: Resistors.

1. Assemble the voltage divider circuit shown in Figure 6.27. Use $V = 4.4$ V and pick two resistors so that $R_1 = R_2$, with the current in the loop being 1 mA. After you pick the resistors, measure their resistance.

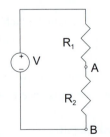

Figure 6.27. Circuit for part 1 of Lab Project 6.5.1.

2. Calculate and measure V_{Thev} and R_{Thev} between points A and B. Compare the calculated and measured values.

3. Suppose we now use V_{AB} as a voltage source to some circuit that has an effective resistance R_L, as shown in Figure 6.28. Using the following values of $R_L = 14\,\text{k}\Omega$, $10\,\text{k}\Omega$, $100\,\text{k}\Omega$, and $1\,\text{M}\Omega$, measure V_{AB} and compare it with your predictions.

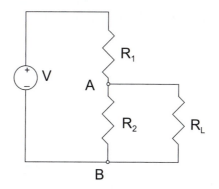

Figure 6.28.   Circuit for part 3 of Lab Project 6.5.1.

4. Repeat the calculation of V_{AB}. Pick the most convenient method. What do you conclude?

6.5.2   AC Signals

Required equipment: Function generator, multimeter, oscilloscope, function generator with ungrounded common.

6.5.3   Diagnosing AC Signals

We have two types of diagnosis tools at our disposal: the multimeter and the scope (short for oscilloscope). By the end of the course, you are expected to master the use of these two measuring tools.

1. Set the function generator to a sinusoidal waveform of amplitude 5 V (that is, 10 V peak to peak) and frequency 1 kHz. Use the scope to verify these settings. Measure the voltage of the function generator using the multimeter. Explain the result.

2. Now add a DC offset of 5 V to the signal. The function generator can do this on its own. Using this input signal, explain the difference between the "DC" and "AC" settings of the scope. (The answer is *not* that "DC" is for measuring DC voltages and "AC" is for measuring AC voltages.)

3. Switch the input signal to square wave. The function generator allows you to change the duty cycle of the square wave. Measure the signal with the multimeter for different values of the duty cycle. What does the multimeter measure?

4. Set the function generator back to sinusoidal output with no zero offset, an amplitude of 2 V, and a frequency of 500 Hz. Assemble a circuit with the function generator as the source connected to a 50 Ω resistor. Explain the value of the voltage that you

measure across the 50 Ω resistor. If you are out of clues, try using a 100 Ω resistor instead.

5. Now use the function generator with ungrounded common. If you do not have one, then skip this part. Assemble the circuit of Figure 6.29 with $V_0 = 2$ V, $R = 1$ kΩ, and $C = 0.1$ μF. Connect channels 1 and 2 of the oscilloscope as shown and invert the channel 2 input (an option of the oscilloscope). Compare and observe the two signals when you change the frequency from 1 kHz to 5 kHz.

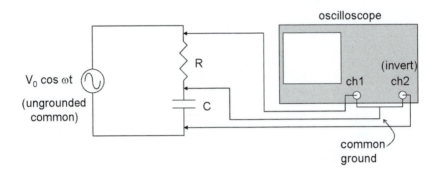

Figure 6.29.   Circuit for Lab Project 6.5.3, on measuring current and voltage together in a circuit.

6.5.4 Impedance Matching

Required equipment: Fast oscilloscope (100 MHz or higher), pulse generator.
Required components: RG-58 coaxial (50 Ω) cable 10 m (25 ft) or longer, cable tee, 50-Ω terminator.

With the pulse generator, generate as short a pulse as possible: 50 ns or shorter is ideal, though it still works with 1–10 μs pulses (but use longer cable). Connect the cable to an oscilloscope and diagnose the signal. Connect the coaxial tee to the cable on one end and to the terminator at the other end, and then plug the tee to the oscilloscope, as shown in Figure 6.30. Diagnose the signal and compare it to the one obtained without the terminator.

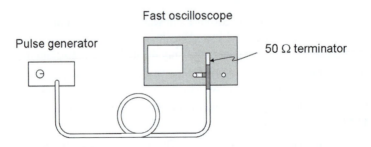

Figure 6.30.   Setup for measuring short pulses, as part of Lab Project 6.5.4.

Another experiment involves using another tee at the source end, connected to a short cable, with the other end also connected to a terminated tee and oscilloscope. Measure the signal propagation time through the cable and calculate the propagation speed.

6.5.5 Practicum Test

You are given a mystery box that puts out a periodic signal.

1. Use the scope to diagnose the signal and make a graph of voltage vs. time. You must label the vertical and horizontal axes. (If you are really stuck, you can ask for help. The amount of help will count against your grade. However, an incomplete circuit gets less credit than a completed one.)

2. You must also determine:

   (a) The amplitude (peak) of the signal
   (b) The DC voltage offset
   (c) The frequency

3. Put together a voltage divider circuit that divides the signal voltage by a factor of two. This should not change significantly when you connect a $5\,\text{k}\Omega$ load resistor to the divider.

CHAPTER 7

FILTERS AND THE FREQUENCY DOMAIN

Contents

So far, we have been focusing on currents and voltages, but AC signals bring a new parameter into electronics: frequency. Electronic devices may respond differently depending on the frequency of the input signal. Moreover, an electronic signal may be the sum of two or more signals of different frequencies. The obvious example is music. Audio signals are a mixture of frequencies, to the extreme that a single-frequency signal, a pure tone, is the least common. An essential element of any audio system is the control of low and high frequencies. Basic audio systems control the amount of high and low frequencies, but a sophisticated audio system has an equalizer, which controls the relative amount of several ranges of frequencies. The output of the audio system is also divided, with high tones sent to one type of speaker (the tweeter), and low tones sent to a different one (the woofer).

How do we provide all these frequency discriminations of electronic signals? We do this with filters. Audio is just one important example, but frequency filtering is important in numerous other situations, so filters constitute an important type of AC circuit. Filters allow the selection or rejection of some AC frequencies over others to serve some specific purpose. In the previous chapter, we saw that capacitors and inductors have impedances that depend on the frequency of the input signal. Thus, they provide us with the building blocks of filters. In this chapter we discuss only the simplest types of filters. It is not a comprehensive discussion, but it covers a few basic circuits and gives you the know-how to dig in deeper if you want to go further.

7.1 RC FILTERS

We start by using capacitors as our frequency-dependent elements. In RC filters, the other element is a resistor. These are the simplest and most common type of filter. They can be understood using the voltage divider concept. Suppose we have an input voltage applied to a voltage divider, as shown in Figure 7.1. The elements of the voltage divider may have frequency dependencies specified by their impedance. The output voltage is then given by

$$V_{\text{out}}(\omega) = \frac{z_2(\omega)}{z_1(\omega) + z_2(\omega)} V_{\text{in}} \tag{7.1}$$

In the case of RC filters, this basic circuit leaves us with two options, or types of RC filters: high-pass and low-pass. In the following sections we discuss these two separately and in detail. One further note is needed before we proceed: Notice the way we have set up the previous circuit. It is not a circuit in itself, but rather a circuit that takes an input signal and delivers it to a circuit that follows. The more we view circuits in this modular way, the more we will be able to quickly understand complex circuits.

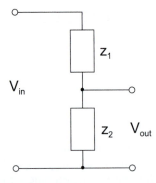

Figure 7.1.   A filter is a voltage divider with frequency-dependent components.

7.2 HIGH-PASS FILTERS

Figure 7.2 shows the RC high-pass filter. We can see roughly how it works by analyzing its behavior at the two extremes of the driving frequency ω. When $\omega = 0$, the capacitor acts as an open circuit so that no DC current is flowing. In this particular case, the capacitor is called a *blocking* capacitor, because it prevents DC signals from proceeding through the circuit. With no DC current, there is no voltage drop across the resistor, so $V_{out} = 0$. At the other frequency limit, when $\omega \to \infty$, the capacitor acts as a short; consequently all the input voltage appears across the resistor, and $V_{out} = V_{in}$.

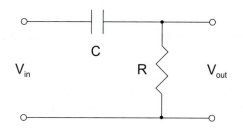

Figure 7.2. RC high-pass filter.

Using the voltage divider concept and the impedances of the two elements, we can get an analytical expression for the output voltage:

$$V_{out} = \frac{R}{R - j/(\omega C)} V_{in} \tag{7.2}$$

This equation can be simplified using the exponential notation of complex numbers discussed in the previous chapter:

$$V_{out} = \frac{R}{\sqrt{R^2 + 1/(\omega C)^2}} V_{in} e^{-j\theta_1} \tag{7.3}$$

where

$$\theta_1 = \tan^{-1}\left(-\frac{1}{\omega RC}\right) \tag{7.4}$$

The term that multiplies V_{in} in Equation 7.2 is also known as the *transfer function*. Note also that the denominator is a function of ω. It is zero when $\omega = j/(RC)$. This is the "pole" of the filter. In general, the transfer functions of filters come down to an expression with a polynomial in the denominator. The roots of the polynomial are the poles. The amplitude of V_{out} is

$$|V_{out}| = \frac{R}{\sqrt{R^2 + 1/(\omega C)^2}} |V_{in}| \tag{7.5}$$

We can define the transmission amplitude of the filter as follows:

$$T = \frac{|V_{out}|}{|V_{in}|} \tag{7.6}$$

Then the transmission amplitude of a signal at frequency ω is

$$T = \frac{R\omega C}{\sqrt{1 + R^2 \omega^2 C^2}} \tag{7.7}$$

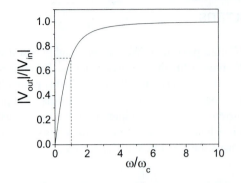

Figure 7.3.   Transfer function of the RC high-pass filter.

As discussed earlier, you can see from the expression that $|V_{out}| \to 0$ when $\omega \to 0$, and $|V_{out}| \to |V_{in}|$ when $\omega \to \infty$. Figure 7.3 shows a plot of T vs. ω. The phase varies from $\theta_1 = -\pi/2$ when $\omega = 0$, to $\theta_1 \to 0$ when $\omega \to \infty$. An important point in the curve is the cutoff angular frequency:

$$\omega_c = \frac{1}{RC} \tag{7.8}$$

It is defined as the point where the ratio of the amplitudes of the two voltages is $1/\sqrt{2} = 0.707$. It also marks the point where the amplitude starts to drop appreciably as $\omega \to 0$. Using Equation 7.8, we can rewrite Equation 7.7 as

$$T = \frac{(\omega/\omega_c)}{\sqrt{1 + (\omega/\omega_c)^2}} \tag{7.9}$$

In electrical engineering, the filter curve is normally expressed on a log-log scale, as shown next. It is also known as a Bode plot. The scale is given such that the values of T and ω/ω_c can be read directly off the graph. That is, the physical length of the horizontal axis is logarithmically dependent on ω/ω_c but linearly dependent on $\log(\omega/\omega_c)$. Next we analyze the main features of that curve. If we take the logarithm (base 10) of T in Equation 7.9, we get

$$\log T = \log \frac{\omega/\omega_c}{\sqrt{1 + (\omega/\omega_c)^2}} \tag{7.10}$$

We can see that when $\omega \ll \omega_c$, this is true:

$$\log T \simeq \log \frac{\omega}{\omega_c} \tag{7.11}$$

This means that for $\omega/\omega_c < 1$, the graph of the $\log T$ is linear with $\log(\omega/\omega_c)$. We can appreciate this by the nearly straight slope in the early part of the graph in Figure 7.4. In this type of graph, we can easily characterize the steepness of the attenuation by the slope of the curve. Every time the frequency doubles (or increases by an octave), the log of the relative voltage increases by a factor of $\log 2$ or 6 dB, so this slope is also referred to as a 6-dB-per-octave slope. This fall of the transmission curve is also known as the frequency roll-off, with the region of suppressed frequencies known as the stop-band, and the region of transmitted frequencies known as the passband. The "knee" of the curve is at the cut-off frequency. Because at this frequency the ratio of voltage amplitudes is $1/\sqrt{2}$ or 3 dB (see Table 6.1), the cutoff frequency is also called the 3 dB point.

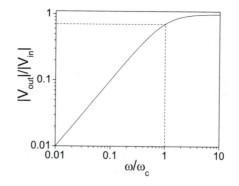

Figure 7.4. Bode plot of the transfer function of the RC high-pass filter.

7.3 LOW-PASS FILTER

The circuit for a low-pass filter is the same as for a high-pass filter but with the components' positions interchanged, as shown in the circuit of Figure 7.5. Using the voltage divider argument, we get

$$V_{\text{out}} = \frac{-j/(\omega C)}{R - j/(\omega C)} V_{\text{in}} \tag{7.12}$$

or

$$V_{\text{out}} = \frac{1/(\omega C)}{\sqrt{R^2 + 1/(\omega^2 C^2)}} V_{\text{in}} e^{-j(\pi/2 + \theta_1)} \tag{7.13}$$

where θ_1 is also defined by Equation 7.4. The ratio of the amplitudes is

$$\frac{|V_{\text{out}}|}{|V_{\text{in}}|} = \frac{1}{\sqrt{1 + R^2 \omega^2 C^2}} \tag{7.14}$$

or

$$T = \frac{1}{\sqrt{1 + (\omega/\omega_c)^2}} \tag{7.15}$$

where the cutoff frequency here is also given by $\omega_c = 1/RC$. Figure 7.6 shows the graph of the transmission curve. Notice that it is not a sharp cutoff. We can also see that

$$\log T = \log \frac{1}{\sqrt{1 + (\omega/\omega_c)^2}} \tag{7.16}$$

When $\omega \gg \omega_c$, we get

$$\log T \simeq -\log \frac{\omega}{\omega_c} \tag{7.17}$$

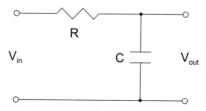

Figure 7.5. RC low-pass filter.

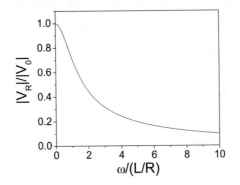

Figure 7.6.   Transfer function of the RC low-pass filter.

The log-log graph has a frequency roll-off of $-6\,\text{dB}$ per octave for $\omega/\omega_c > 1$. Figure 7.7 shows the graph of the filter curve in a log-log scale.

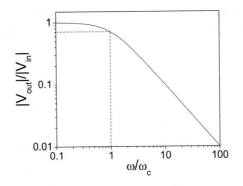

Figure 7.7.   Bode plot of the transfer function of an RC low-pass filter.

7.4   CASCADING FILTERS

In the filter curves shown in the two previous parts, we can see that the amplitude of the output decreases slowly. We may want to have a filter curve that decreases much more sharply. Consider the result of having n filters connected in series, as shown in Figure 7.8.

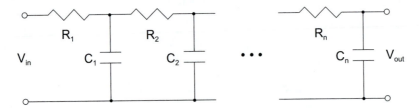

Figure 7.8.   Cascading filter sections.

If we make $\omega_{c1} = \omega_{c2} = \cdots = \omega_c$, the transfer function becomes:

$$T = \left[\frac{1}{\sqrt{(\omega/\omega_c)^2 + 1}} \right]^n \tag{7.18}$$

When $\omega \gg \omega_c$, then

$$\log T \simeq -n \log \frac{\omega}{\omega_c} \qquad (7.19)$$

resulting in a sharper roll-off of $-6n$ dB per octave, as shown in the graph of Figure 7.9 for $n = 2$.

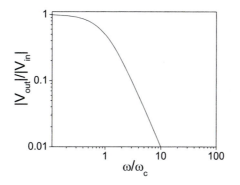

Figure 7.9. Bode plot of the transfer function of a two-stage low-pass filter.

■ **EXERCISE 7.1**

The -3 dB frequency of the n-stage filter is not ω_c. Find an expression for the -3 dB frequency as a function of n and ω_c.

A filter with n stages has a transfer function that has a polynomial of order n in the denominator, and so it has n poles. The higher the order of the filter, the sharper the frequency roll-off. However, we cannot put identical filters cascaded together, because of impedance mismatches. That is, a filter stage may load the previous stage too much. As you recall, when a circuit gets loaded, V_{out} will be much less than V_{in}. Thus, if we simply connect two identical filters in series, even the passing frequency will be attenuated. We can overcome this by having cascaded filters with the same cutoff frequencies but with component values such that each stage does not load the previous stage.

Let us analyze the case of two cascaded filters ($n = 2$). To analyze the loading effect, we have to first find the Thevenin equivalent circuit of the first stage (Figure 7.10).

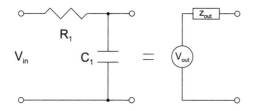

Figure 7.10. Thevenin equivalent of a filter stage.

The Thevenin voltage is just the filter voltage, of amplitude

$$V_{out} = \frac{V_0}{\sqrt{1 + R^2 \omega^2 C^2}} \qquad (7.20)$$

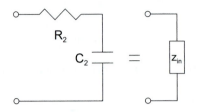

Figure 7.11.   Input impedance of filter stage.

The Thevenin resistance, also known as the *output impedance* for this case, is the resistor and capacitor in parallel. The output impedance is then

$$\frac{1}{Z_{\text{out}}} = \frac{1}{R_1} + j\omega C_1 \tag{7.21}$$

which has an amplitude of

$$|Z_{\text{out}}| = \frac{R_1}{\sqrt{1 + R_1^2\omega^2 C_1^2}} \tag{7.22}$$

We can see that the minimum value of $|Z_{\text{out}}|$ is zero for $\omega \to \infty$, and its maximum value is R_1 for $\omega \to 0$. We can think of the second stage as a load with impedance Z_{in} (Figure 7.11). This second filter stage can be represented by an equivalent impedance, called the *input impedance*, given by the impedance of the two elements in series:

$$Z_{\text{in}} = R_2 - \frac{j}{\omega C_2} \tag{7.23}$$

This impedance has an amplitude

$$|Z_{\text{in}}| = \sqrt{R_2^2 + 1/(\omega C_2)^2} \tag{7.24}$$

The maximum impedance is infinite for $\omega \to 0$. Its lowest value is R_2 when $\omega \to \infty$. If we connect both, for the second stage not to load the first stage, we need $|Z_{\text{out}}| \ll |Z_{\text{in}}|$. It can be shown that

$$\frac{|Z_{\text{out}}|}{|Z_{\text{in}}|} = \frac{R_1}{R_2\sqrt{2 + \omega^2/\omega_c^2 + \omega_c^2/\omega^2}} \tag{7.25}$$

This is zero for $\omega \to 0$ and $\omega \to \infty$, and it has a maximum value of $R_1/(2R_2)$ for $\omega = \omega_c$. Therefore, a good rule of thumb is to use $R_2 = 10R_1$. To have both filters with the same cutoff frequency, we make $C_2 = C_1/10$.

We are still at a disadvantage when we connect the output of the cascaded filter to a low-impedance load. For example, connecting the output of an audio filter to an 8-Ω speaker may load the filter completely. The solution to this problem is to use an impedance transformer, discussed later.

7.5 IMPORTANT CONSIDERATIONS FOR FILTER DESIGN

Before we continue, we need to stop and consider some design issues.

7.5.1 f vs. ω

The equations describing the previous two filters use *angular* frequencies ω and $\omega_c = 1/RC$. Their units are radians per second, or rad/s. When using a function generator, we talk in terms of the frequency $f = \omega/2\pi$, which has as units cycles per second, or Hz. Care must be exercised to use the correct frequency when designing filters. Confusing f with ω can lead to incorrectly designed filters.

7.5.2 Determining ω_c

It is important to realize that these filters are not ideal. The cutoff is not sharp. Consider the following example: Suppose that we want to get rid of an unwanted frequency at 30 kHz while transmitting lower frequencies. First, we conclude that we need a low-pass filter. How many stages? To answer this question, we need to get more specific about how much of that frequency we want to eliminate. Let us consider a 10-percent rule (that is, aiming for $T = 0.1$ at 30 kHz). Let us consider a one-stage filter. If we pick $f_c = 30$ kHz, then from Equation 7.15 we get $T = 1/\sqrt{2} = 0.71$. That is, the output amplitude will be 71 percent of the input amplitude, which is not good. If we naively set $f_c = 0.9 f$ (the cutoff frequency is ten percent less than the undesired frequency), we get $T = 0.67$. Not good either! The problem is that the filter function is not very sharp. Turning Equation 7.15 around, if we want $T = 0.1$ at frequency f (30 kHz), then we calculate that $f_c \sim f/10 = 3$ kHz. With this filter we also get $T \geq 0.9$ for frequencies below 1.45 kHz. If that is okay, then we are done, and use 3 kHz as the cutoff frequency. If, on the other hand, we want $T \geq 0.9$ at a higher frequency, such as 3 kHz, then we have a problem: This filter gives us only $T = 0.71$ when $f_c = 3$ kHz. To resolve this situation we need a sharper filter curve. If we use two stages, then by using Equation 7.18 with $n = 2$, we get that the cutoff for $T = 0.1$ at $f = 30$ kHz is now $f_c = 10$ kHz, for which case $T = 0.92$ at $f = 3$ kHz. Therefore, we must design filters by first deciding the desired values of T at the frequencies of interest.

7.6 TRANSFORMER

In this course we cover three types of impedance transformers: an inductive transformer, a transistor-based emitter-follower, and an operational amplifier buffer. We discuss the first one now and cover the second and third ones in Chapters 9 (Transistors) and 10 (Operational Amplifiers), respectively. A transformer, shown in Figure 7.12, is a system of two coils that

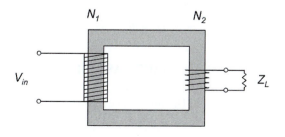

Figure 7.12. A transformer.

share the same magnetic field through an iron core. For this brief discussion we assume that we have a transformer with no losses. Thus, both the primary and the secondary share the same magnetic flux. From the laws of induction, which connect the voltages to the rate of change of the magnetic flux, we get a relationship between the voltages of the primary $V_1 = V_{in}$ and the secondary V_2. If the primary and the secondary have N_1 and N_2 turns, respectively, the relationship between the primary and secondary voltages is

$$\frac{V_2}{V_1} = \frac{N_2}{N_1} \tag{7.26}$$

At this point we must indulge in an important tangent related to Equation 7.26. Transformers are the workhorse of the AC industry via that equation. With it, we can boost a voltage or reduce it. When we plug a piece of equipment into the wall outlet, it receives an AC signal with a given RMS value, typically 110 V. If the device that plugs into it uses high voltages (for example, TV, microwave, vacuum tubes, specialty discharge lamps, or high-voltage supplies), then the first step in the internal circuitry is to boost the line voltage with a transformer with $N_2 > N_1$. These high voltages make the internal circuitry particularly dangerous because of the risk of injury by electrocution, hence the warnings to not operate it without the cover. If the device that we plug into the wall uses low voltages (such as radios and other consumer electronics), we must reduce the line voltage via $N_2 < N_1$. In the next chapter we discuss how to proceed further to produce the DC voltages that digital and other circuits need. End of tangent.

If we assume no losses, then all the power in the primary is transferred to the secondary, or $V_1 I_1 = V_2 I_2$. If we combine this with Equation 7.26, we get a relationship between the currents:

$$\frac{I_2}{I_1} = \frac{N_1}{N_2} \tag{7.27}$$

If we have $N_1 > N_2$, we can boost the current. One direct application of this is an arc welder, in which the large current creates a high enough temperature to melt and fuse metals.

If we connect an impedance Z_L to the secondary, the transformer makes this impedance look to the primary as

$$Z_{in} = \frac{V_1}{I_1} = Z_L \left(\frac{N_1}{N_2}\right)^2 \tag{7.28}$$

thereby acting as an impedance transformer. For example, an audio speaker has an impedance of 8 Ω. This is too low for the output impedance of audio generators or filters. If we put a transformer with a 10:1 ratio of windings between the speaker and the source, the speaker will appear to the source as an 800-Ω load.

■ **EXERCISE 7.2**

Derive Equation 7.28 using $Z_{in} = V_1/I_1$ and $V_2/I_2 = Z_L$.

7.7   RESONANT CIRCUITS AND BAND-PASS FILTERS

Consider the circuit of Figure 7.13. The impedance between points A and B, Z_{LCP}, is

$$Z_{LCP} = \left(\frac{1}{j\omega L} + j\omega C\right)^{-1} \tag{7.29}$$

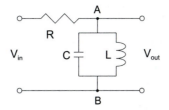

Figure 7.13.   Resonant LC filter.

After some rearranging, we get

$$Z_{LCP} = \frac{-j\omega L}{\omega^2 LC - 1} \tag{7.30}$$

Since the circuit can be represented as a voltage divider, as shown in Figure 7.14, then

$$V_{out} = \frac{Z_{LCP}}{R + Z_{LCP}} V_{in} \tag{7.31}$$

We can see from Equation 7.30 that $Z_{LCP} = \infty$ when $\omega = \omega_0$, where

$$\omega_0 = \frac{1}{\sqrt{LC}} \tag{7.32}$$

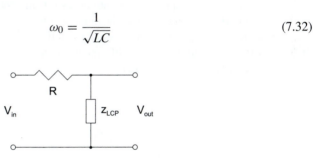

Figure 7.14.   Band-pass LC filter as a voltage divider.

We can also see that $Z_{LCP} = 0$ when $\omega = 0$ or $\omega = \infty$. This means that $|V_{out}| = |V_{in}|$ when the circuit is in resonance ($\omega = \omega_0$). We can see that $|V_{out}|/|V_{in}| \to 0$ as ω gets farther away from ω_0 in either direction; the circuit acts as a *band-pass filter*. More formally,

$$T = \frac{|V_{out}|}{|V_{in}|} = \frac{\omega L}{\sqrt{R^2(\omega^2 LC - 1)^2 + \omega^2 L^2}} \tag{7.33}$$

with the curve shown in Figure 7.15.

The sharpness of the curve is defined by the quality factor Q of the circuit, given by

$$Q = \frac{\omega_0}{\Delta\omega} \tag{7.34}$$

where $\Delta\omega$ is the frequency difference between the two 3 dB points at either side of the curve. In terms of the circuit components, Q can be expressed as

$$Q = \omega_0 RC = \frac{R}{\omega_0 L} \tag{7.35}$$

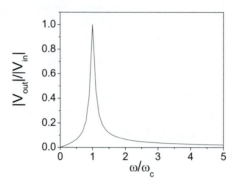

Figure 7.15.   Transfer function of the resonant LC band-pass filter.

The filter curve can now be reformulated to include Q as

$$T = \frac{\omega/\omega_0}{\sqrt{Q^2[(\omega/\omega_0)^2 - 1]^2 + (\omega/\omega_0)^2}} \tag{7.36}$$

Figure 7.16 gives a graph of resonant curves for different values of Q. Note that this analysis ignores the resistance of both the inductor and the load, which, in a practical circuit, must be included. A large inductor resistance also limits the Q of the filter, but we have not included it in this discussion to simplify the calculations. Inductor resistances are always small, but they become a factor when we want a very high Q.

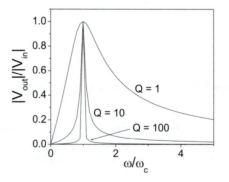

Figure 7.16.   Transfer functions of the resonant LC band-pass filter as a function of Q.

This parallel circuit is used in radio receivers. Figure 7.17 gives a simplified schematic of a radio receiver. Notice that it has a transformer to couple the signal from the antenna into the LC circuit. The antenna receives elecromagnetic waves of different frequencies (radio stations; AM radio is from 540 kHz to 1600 kHz) and couples them into the circuit. The secondary of the transformer is the inductor in the LC circuit. By varying the value of the capacitor, we decide the value of the frequency that is resonant with the circuit. The tuner of the radio is nothing other than a variable capacitor. After the filter, the resonant signal is then rectified and demodulated because the actual information (music) is encoded in the modulation of the carrier frequency. In AM radio the amplitude is modulated, whereas in FM radio it is the frequency that is modulated.

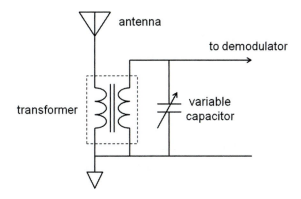

Figure 7.17. Simplified schematic of a radio receiver circuit.

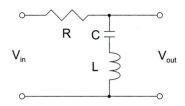

Figure 7.18. Resonant LC notch filter.

Next, we consider an alternate arrangement of components, shown in Figure 7.18. This is a filter with a capacitor and an inductor in series. It can be shown that for this case

$$T = \frac{\omega^2 LC - 1}{\sqrt{\omega^2 R^2 C^2 + (\omega^2 LC - 1)^2}} \tag{7.37}$$

which is also resonant at $\omega = \omega_0 = 1/\sqrt{LC}$. This is also known as a notch filter, which is designed to suppress an unwanted frequency ω_0. When the Q of the filter is given by $Q = \omega_0 L/R = 1/(\omega_0 RC)$, then the filter equation becomes

$$T = \frac{|(\omega/\omega_0)^2 - 1|}{\sqrt{[(\omega/\omega_0)^2 - 1]^2 + [\omega/(\omega_0 Q)]^2}} \tag{7.38}$$

When the inductor has a finite resistance, the dip of the notch does not go all the way down to zero, but goes to $(R_{inductor})/(R + R_{inductor})$. The inductor resistance also affects the Q of the filter. A plot of filter curves for different values of Q is shown in Figure 7.19.

7.8 HIGHER-ORDER FILTERS

The design of filters is a mature topic in electronics. You can imagine all the places where filters are used in a fundamental way; audio systems and telephone communications are only two examples. A number of filter designs offer many features, among them sharp transmission curves. To accomplish this, you need to combine capacitors with inductors and use several stages. Among the classic filters are the Butterworth, Chebyshev, and Bessel designs. We do not devote further coverage of these, but we revisit filters in

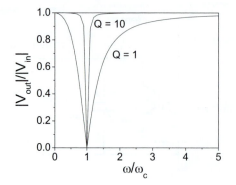

Figure 7.19.   Transfer function of the resonant LC notch filter.

Chapter 10 (Operational Amplifiers), where the use of feedback and amplification leads to *active* filters. These filters reduce the complexity of *passive* filters, the kind discussed in this chapter, and easily overcome the latter's impedance-matching shortcomings.

7.9 FOURIER SERIES

A remarkable theorem proposed by Joseph Fourier in the early 1800s applies to nature's behavior in remarkably many different situations. In electronics, this particular point of view gives us great insight into processing complex signals. Soon we will be moving from sinusoidal signals to more complex ones, and we need a way to adjust our analysis accordingly. Analyzing complex signals can be reduced to the sinusoidal analysis done so far by applying the Fourier theorem.

If a function $f(t)$ is a bounded periodic function of period T (that is, $f(t + T) = f(t)$) and satisfies the Dirichlet conditions:

1. In any period, $f(t)$ is continuous, except for a finite number of jump discontinuities.

2. In any period, $f(t)$ has only a finite number of maxima and minima.

then $f(t)$ can be represented by the Fourier series

$$f(t) = \sum_{n=0}^{\infty} \left[a_n \sin\left(\frac{2n\pi t}{T}\right) + b_n \cos\left(\frac{2n\pi t}{T}\right) \right] \tag{7.39}$$

where a_n and b_n are determined by

$$a_n = \frac{2}{T} \int_{-T/2}^{+T/2} f(t) \sin\left(\frac{2n\pi t}{T}\right) dt \tag{7.40}$$

and

$$b_n = \frac{2}{T} \int_{-T/2}^{+T/2} f(t) \cos\left(\frac{2n\pi t}{T}\right) dt \tag{7.41}$$

for $n = 1, 2, \ldots$, and with

$$a_0 = 0 \tag{7.42}$$

f(t)

Figure 7.20.   Peridic signal as a function of time.

and

$$b_0 = \frac{1}{T} \int_{-T/2}^{+T/2} f(t)dt \tag{7.43}$$

This means that *any* periodic signal of frequency $\omega_0 = 2\pi/T$ (see for example, Figure 7.20) can be decomposed into a sum of sines and cosines that are harmonics of the fundamental frequency.

In electronics, this finding may be applied to analyzing the effect of a circuit on a complex signal, by considering its effect on the individual frequency components of the input signal.

The previous coefficients can also be written as

$$a_n = \frac{2}{T} \int_{\tau-T/2}^{\tau+T/2} f(t) \sin\left(\frac{2n\pi t}{T}\right) dt \tag{7.44}$$

$$b_n = \frac{2}{T} \int_{\tau-T/2}^{\tau+T/2} f(t) \cos\left(\frac{2n\pi t}{T}\right) dt \tag{7.45}$$

and

$$b_0 = \frac{1}{T} \int_{\tau-T/2}^{\tau+T/2} f(t)dt \tag{7.46}$$

where τ is any real number. That is, we could have done the integrals of the previous example between any two time limits, as long as they are one period apart.

In addition, the previous representation gets simplified when:

1. $f(t)$ is even (that is, $f(-t) = f(t)$), which results in $a_n = 0$ ($n \neq 0$).
2. $f(t)$ is odd (that is, $f(-t) = -f(t)$), in which case $b_n = 0$.

As will be shown in the examples that follow, b_0 represents the DC component of the signal. That is, if its time average is zero, then $b_0 = 0$.

Let us do a first example: The case of a square wave shown in Figure 7.21. We can immediately see that it has a DC offset of $V_0/2$ and that it is an odd function. The function can be defined as:

$$V(t) = \begin{cases} V_0 & 0 < t \leq T/2 \\ 0 & T/2 < t \leq T \end{cases}$$

Applying Equation 7.43, we get

$$b_0 = \frac{1}{T} \int_0^{T/2} V_0 dt = \frac{V_0}{2}$$

as we deduced from just inspecting the waveform.

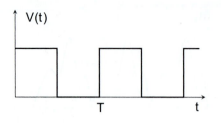

Figure 7.21.   Square wave.

■ **EXERCISE 7.3**

Sketch the square wave that would give $b_0 = -V_0/2$.

The coefficients for the cosine terms of the series are

$$b_n = \frac{2}{T} \int_0^{+T/2} V_0 \cos\left(\frac{2n\pi t}{T}\right) dt$$

$$= \frac{V_0}{n\pi} \left[\sin\left(\frac{2n\pi t}{T}\right)\right]_0^{T/2}$$

$$= 0$$

which makes sense because the function is odd. The coefficients for the sine terms are

$$a_n = \frac{2}{T} \int_0^{+T/2} V_0 \sin\left(\frac{2n\pi t}{T}\right) dt$$

$$= \frac{V_0}{n\pi} \left[-\cos\left(\frac{2n\pi t}{T}\right)\right]_0^{T/2}$$

$$= \frac{V_0}{n\pi}[-\cos(n\pi) + 1]$$

$$= \frac{2V_0}{n\pi}$$

for n odd; otherwise, they are zero.

■ **EXERCISE 7.4**

Sketch on the previous graph of the square wave:

1. $\frac{2V_0}{\pi} \sin\left(\frac{2\pi t}{T}\right)$
2. $\frac{2V_0}{3\pi} \sin\left(\frac{6\pi t}{T}\right)$
3. $\frac{2V_0}{2\pi} \sin\left(\frac{4\pi t}{T}\right)$
4. Use those sketches to explain why terms (1) and (2) are part of the Fourier decomposition and not term (3).

The square wave can then be expressed analytically as

$$V(t) = \frac{V_0}{2} + \sum_{n=1}^{\infty} \frac{2V_0}{n\pi} \sin\left(\frac{2n\pi t}{T}\right)$$

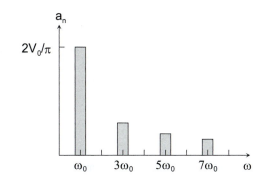

Figure 7.22.   Frequency spectrum of a square wave.

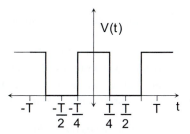

Figure 7.23.   Symmetric square wave.

(n odd). Notice that here we have applied Equations 7.44, 7.45, and 7.46 conveniently for $\tau = T/2$. Figure 7.22 gives the frequency spectrum of the square wave.

Let us redo the calculation for a symmetric square wave, as shown in Figure 7.23. The function can be expressed as

$$V(t) = \begin{cases} V_0 & -T/4 < t \le T/4 \\ 0 & T/4 < |t| \le T/2 \end{cases}$$

Just by inspection, we see that it has the same DC offset as in the previous example, but note that it is an even function (symmetric about $t = 0$). The coefficients are calculated as

$$b_0 = \frac{1}{T} \int_{-T/4}^{+T/4} V_0 dt$$

$$= \frac{V_0}{2}$$

$$a_n = \frac{2}{T} \int_{-T/4}^{+T/4} V_0 \sin\left(\frac{2n\pi t}{T}\right) dt$$

$$= \frac{V_0}{n\pi} \left[-\cos\left(\frac{2n\pi t}{T}\right) \right]_{-T/4}^{T/4}$$

$$= 0$$

and

$$b_n = \frac{2}{T} \int_{-T/4}^{+T/4} V_0 \cos\left(\frac{2n\pi t}{T}\right) dt$$

$$= \frac{V_0}{n\pi} \left[\sin\left(\frac{2n\pi t}{T}\right)\right]_0^{T/2}$$

$$= \frac{2V_0}{n\pi} \sin\left(\frac{n\pi}{2}\right)$$

which is different than zero for n odd. The resulting analytical expression is

$$V(t) = \frac{V_0}{2} + \sum_{n=1}^{\infty} \frac{2V_0}{n\pi} \sin\frac{n\pi}{2} \cos\left(\frac{2n\pi t}{T}\right)$$

(n odd). Notice that the frequency components and their amplitude have not changed. Only the phase of the function has changed.

The first and second examples are cases of odd and even functions, respectively. We found that their Fourier decomposition was in terms of either sines or cosines, depending on the parity of the function. In general, a wave with an arbitrary phase has both sines and cosines in its Fourier decomposition. For example, consider the following signal:

$$V(t) = V_0 \cos(\omega t + \phi)$$

We can show that its Fourier terms reduce to:

$$a_n = a_1 = \cos\phi$$

and

$$b_n = b_1 = -\sin\phi$$

If a function is more complex than a simple harmonic, its Fourier decomposition has higher-order terms; the sharper the slopes and kinks in the function, the higher the harmonics involved. In electronics, Fourier decomposition is important because we can analyze the processing of an electronic signal by studying the effect of the electronic system (filters, amplifiers) on the individual frequency components of the signal.

7.10 PROBLEMS

1. For the filter of Figure 7.24:

   (a) Identify the type of filter.

   (b) Find the cutoff frequency.

   (c) Attach a second stage with the same cutoff frequency that does not load the first one.

2. Design an RC low-pass filter with a cutoff frequency at 2 kHz, using a capacitor of 0.1 μF.

   (a) Show a circuit diagram, and your calculations in obtaining R.

   (b) Make a sketch of the absolute value of the input impedance $|Z_{in}|$ of the filter vs. frequency. What are the limiting values of $|Z_{in}|$?

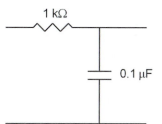

Figure 7.24. Circuit for Problem 1.

(c) Make a similar sketch of the output impedance of the filter.

(d) If we connect the filter to a load resistance R_L, what minimum value should it have in order to not load the filter?

(e) Design two more filter stages to sharpen the filter curve (with the same cutoff frequency). Your design should address the problem of loading. Make a log-log plot of the filter curve using the software of your preference.

3. An audio source has an output impedance of 200 Ω. Design a filter that transmits more than 90% of frequencies above 15 kHz, but less than 10% of frequencies below 2 kHz. Draw a complete circuit diagram, including resistor and capacitor values.

4. Design a band-pass filter with the following specifications.

(a) Its first stage is a high-pass filter with a cutoff frequency of 1 kHz. The input impedance of the first stage should be at least 500 Ω.

(b) A second stage that follows should be a low-pass filter with a cutoff frequency of 10 kHz. The second stage should not load the first stage.

(c) If we use a transformer to impedance-match the filter to a 10-Ω load, find a ratio of primary to secondary windings in the transformer that will do the job.

5. A 12-V DC power supply that gets its power from the 110-V (RMS) wall outlet needs a transformer as a first step in generating its output. The rectification and regulation stage of the power supply needs an input voltage of 15 V (peak).

(a) What winding ratio should the transformer have?

(b) If the power supply has an effective impedance of 100 Ω (the load impedance that the transformer sees), what impedance does the primary of the transformer see?

6. Design a series LRC filter for 10 kHz using a 0.01 μF capacitor.

(a) What is the Q of the filter?

(b) Redesign the filter with a new Q that is 10 times larger.

(c) Make a log-log plot of the two filter curves vs. frequency.

7. The circuit of Figure 7.25 is supposed to be a band-pass filter, but it has design flaws.

(a) Identify the problems (more than one).

(b) Correct the problems, but keep the same cutoff frequencies.

8. The circuit of Figure 7.26 is a one-stage Butterworth filter. The load is also shown.

(a) Find an analytical expression for the voltage on the load as a function of the frequency and the input voltage.

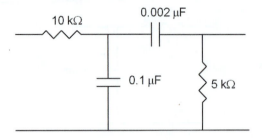

Figure 7.25.   Circuit for Problem 7.

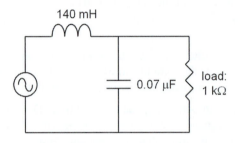

Figure 7.26.   Circuit for Problem 8.

   (b) Make a graph of the filter curve with your favorite software.

   (c) Describe the type of filter and find a cutoff frequency.

   (d) Graph an equivalent RC filter curve and compare it to the previous graph.

9. Calculate the Fourier coefficients of a square wave defined by:

$$V(t) = \begin{cases} V_0/2 & -T/4 < t \le T/4 \\ -V_0/2 & T/4 < |t| \le T/2 \end{cases}$$

10. Calculate the Fourier coefficients for a triangular wave defined by:

$$V(t) = \begin{cases} V_0(1 - 2t/T) & 0 \le t \le T/2 \\ V_0(1 + 2t/T) & -T/2 \le t \le 0 \end{cases}$$

11. For the previous problems, make plots using your favorite software of the first term ($n = 1$) and the sum of the first five terms vs. t. Assume that $T = V_0 = 1$.

12. Suppose that we connect a function generator, set to 1 kHz square waveform with an amplitude of 1 V, to an RC low-pass filter with a frequency cutoff of 2 kHz. Calculate the absolute value of the first five Fourier components (a_n or b_n) of the waveform at the output of the filter.

7.11   LAB PROJECTS

This section presents ideas for lab projects.

7.11.1 Filters

Required equipment: Function generator, AC source, 6.3 VAC vacuum tube-filament supply or the low-voltage (8–12 V-peak) output of a transformer (it should not have a grounded output), oscilloscope, multimeter.
Required components: Resistors, capacitors.

Low-Pass Filter

1. Assemble the circuit of Figure 7.27 for $R = 15\,\mathrm{k\Omega}$ and $C = 0.02\,\mu\mathrm{F}$. Use the function generator as the source with a sine waveform output.

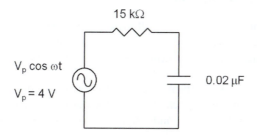

Figure 7.27.   Circuit for low-pass filter lab project.

By making measurements with the scope, obtain a rough value for the cutoff frequency of the filter ($v_c = v_{3dB} = \omega_c/2\pi = 1/(2\pi RC)$). Then take measurements with a multimeter in the AC volts range to reconstruct the filter curve (that is, measurements of V_{out} for a minimum of 10 frequencies). However, for frequencies above a few hundred hertz, the multimeter reading drops due to its internal bandwidth, so above 500 Hz, you should measure both the input and the output and use the ratio for your measurements. Make a regular plot and a Bode plot of the measurements.

2. Connect the ungrounded AC supply in the circuit shown in Figure 7.28. Because it is floating, we can connect it to the ungrounded end of the function generator.

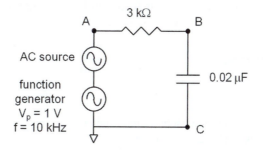

Figure 7.28.   Circuit for second part of low-pass filter lab.

Set the function generator to a sine waveform at a frequency of 10 kHz. Compare V_{AC} with V_{BC} and write your conclusions. Make a sketch of the input and output waveforms. Try varying some of the parameters (such as frequency, waveform) of the function generator and explain what you observe.

Note that you *cannot* measure V_{AB} and V_{BC} simultaneously with the two channels of the scope because their grounds are connected together. If you do, it will short the capacitor or the source. You can still do the comparison with the scope, if it has the utility that subtracts the signals from both channels.

High-Pass Filter   Design a high-pass filter with $R = 8.2\,k\Omega$ and a frequency cutoff of about 2 kHz. Pick the closest capacitor from the capacitor rack. Measure f_c using the scope (as you did for the low-pass filter).

7.11.2   Application: Audio Filter

Required equipment: Oscilloscope, multimeter, CD player with CD of serene music superimposed with 15–20 kHz square wave, earphones or passive speaker.
Required components: Audio transformer, resistors, capacitors.

The Problem   You have been given a CD with a defective recording. The CD has music superimposed with an unwanted pitch. Connect the output of the CD player to a speaker and identify the problem. Look at the audio signal with the scope and identify the frequency of this pitch.

The Solution   Design a three-stage RC filter that should be put in between the CD player and the speaker so that the noise is filtered out (see Figure 7.29). The entire circuit has the

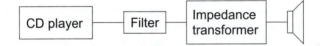

Figure 7.29.   Block diagram for audio filter lab.

four main parts: the CD player (or PC), your filter, an impedance-matching section, and the 8 Ω speaker. At the output of the tape player, you should put a load resistor close to 8 Ω; the player will work best if it sees the 8 Ω. Your three-stage filter must be designed so that each section does not load the previous one (use the $\times 10$ criteria).

Because the output impedance of your filter will be much greater than the speaker's impedance, you must use an impedance-matching device. An audio transformer works well.

7.11.3   Fourier Analysis

Lab credit: J. C. Amato

Required equipment: Function generator, multimeter.
Required components: AD532 multiplier IC.

It is possible to build a simple spectrum analyzer using off-the-shelf integrated circuits. The analyzer you will use today is built around an Analog Devices AD532 multiplier chip. As the name suggests, the output voltage of the device is proportional to the product of the two input voltages. Let us see how the multiplier works in more detail.

Suppose we have two input sinusoidal voltages:

$$V_1(t) = V_{01} \sin \omega_1 t \tag{7.47}$$

and

$$V_2(t) = V_{02} \sin \omega_2 t \tag{7.48}$$

The product of the two voltages is:

$$V_1(t)V_2(t) = V_{01} V_{02} \sin \omega_1 t \sin \omega_2 t \tag{7.49}$$

Using the function-product trigonometric relationship we get

$$V_1(t)V_2(t) = \frac{V_{01} V_{02}}{2}[\cos(\omega_1 - \omega_2)t - \cos(\omega_1 + \omega_2)t] \tag{7.50}$$

If ω_1 and ω_2 are close to each other, then

$$(\omega_1 - \omega_2) \ll (\omega_1 + \omega_2) \tag{7.51}$$

We can use this function multiplier in conjunction with a low-pass filter to measure the Fourier coefficients of an arbitrary periodic function. Let us apply to one of the inputs V_{in} with fundamental frequency ω_0:

$$V_{in} = \sum_{n=0}^{\infty} [a_n \sin(n\omega_0)] \tag{7.52}$$

For simplicity, we have assumed that V_{in} is an odd function. If we multiply V_{in} by a reference waveform

$$V_{ref} = V_0 \sin \omega_r \tag{7.53}$$

where V_0 and ω_r are known, we get the output:

$$V_{out} = \frac{V_0}{2} \sum_{n=0}^{\infty} a_n[\cos(n\omega_0 - \omega_r)t - \cos(n\omega_0 + \omega_r)t] \tag{7.54}$$

We now connect V_{out} to a low-pass filter. If for a given n, where $n = 1, 2, \ldots$, we have $n\omega_0 \sim \omega_r$, then the output of the filter is:

$$V_{out} \simeq \frac{V_0 a_n}{2} \cos(n\omega_0 - \omega_r)t \tag{7.55}$$

Thus, the output is proportional to the Fourier coefficient a_n. If we adjust ω_r to each of the harmonics of ω_0, our output each time will be proportional to the amplitude of the corresponding Fourier component.

1. Assemble the multiplier circuit shown in Figure 7.30. The input V_{ref} should be a sine wave and the V_{in} should be the one whose Fourier components we want to determine. The output (at pin 2) gives $V_{ref} V_{in}/10$.

2. Design a two-stage low-pass filter with a cutoff frequency of 100 Hz and an input impedance of 20 kΩ. Attach it to the output of the multiplier chip.

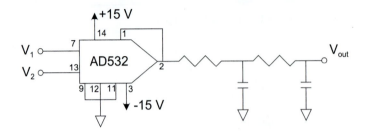

Figure 7.30.   Circuit for measuring the Fourier components of periodic waveform in Lab Project 7.11.3.

3. Apply a 6 V (peak) sine wave of frequency 1 kHz to the V_{ref} input, and a 6 V (peak) 1 kHz square wave to V_{in}. Use the 1 K range in the function generator for the square wave and the 10 K range for the sine wave.

4. Tune V_{ref} (sine wave) through the entire range of frequencies accessible on the 10 K scale. Whenever its frequency ω_r falls close to $n\omega_0$, a sine wave at the difference frequency appears on the scope. Its amplitude is proportional to the Fourier coefficient of the square wave at $n\omega_0$, a_n.

5. Record the amplitude of each Fourier component of the square wave, up to about $n = 21$. Plot the amplitude a_n vs. n on a logarithmic scale. Are the results in agreement with the analysis done in the text?

6. Replace the square wave by a triangular wave, and do the same measurements and analysis. Compare those results with your calculations.

7.11.4 Practicum Test

A mystery box outputs a waveform that you must diagnose and understand.

1. Design and wire a circuit that removes the DC component of the waveform without affecting the AC component.

2. Add a low-pass filter to be placed after the filter stage of the previous part with a cutoff frequency that is 10 times the frequency of the signal.

CHAPTER 8

DIODES

Contents

The electrical components that we have seen so far are *linear*. That is, the current flowing through them is a linear function of the voltage. The evolution of electronics technology has led to the development of nonlinear devices. Some devices started out as vacuum tubes, but solid-state science and technology has replaced vacuum tubes with an assortment of new devices. Nonlinear devices enable us to manipulate electronic signals in interesting and useful ways. We start the discussion of nonlinear devices with diodes.

Diodes have prominent uses today. They mostly play a behind-the-scenes role as rectifiers, but a relatively new application has made them more noticeable: their capability to emit and detect light. These diodes are, respectively, light-emitting diodes (LED) and photodiodes. Laser diodes, a variety of LEDs, also have a prominent role in our everyday life. Yet, whether they are a laser or a rectifier, all diodes behave the same way electronically.

In this chapter we discuss circuit applications that use diodes. We cover in detail their role as rectifiers in DC power supplies, and their new roles as detectors and light emitters. We start our discussion with the physics of semiconductors.

8.1 PHYSICS OF SEMICONDUCTORS

Physicists love simplicity because the laws of nature are best understood when they are put in their simplest forms. Much too often, we see long discussions about a single particle, a single atom, or a single molecule. Physicists often joke that a diatomic molecule has one atom too many. Indeed, even diatomic molecules can be hugely complex to study in detail. If a diatomic molecule is too complex, a solid, with a huge number of atoms, all joined at the hip, goes beyond any simple model. Indeed, the physics of solid-state matter is an entire field of physics by itself. It is rich and marvelously interesting. Contrary to the simple-minded view that it is a mess, solid-state physics has a lot of order and a lot of physics. In this section, we restrict ourselves to probably the most important case: semiconductors. Unfortunately, the discussion cannot be thorough; covering the essentials of semiconductors would require an entire course. Therefore, we discuss just some fundamentals that allow us to grasp the basic physics. This only means a lot of arm-waving arguments—sorry! (For nonphysicists, this refers to the excessive hand gestures that physicists employ to compensate for the lack of rigor in explaining something, to cover for their sense of guilt for skipping so many details, and to convince or distract the skeptical fellow physicist on the receiving end.)

8.1.1 Structure

We start with the ordering of the periodic table of elements, created by Dimitri Mendeleyev in the late nineteenth century. It is ordered into columns of elements that have the same chemical properties. Column IV in the section of the periodic table, shown in Figure 8.1, is particularly important. The elements in this column are tetravalent. That is, they share four electrons with neighboring atoms via covalent bonds. They form a unique crystal structure when all four bonds are equally symmetric. This is the diamond structure, a configuration

	III	IV	V	VI
	B	C	N	O
II	Al	Si	P	S
Zn	Ga	Ge	As	Se
Cd	In	Sn	Sb	Te
Hg	Tl	Pb	Bi	Po

Figure 8.1. Section of the periodic table with semiconductor elements.

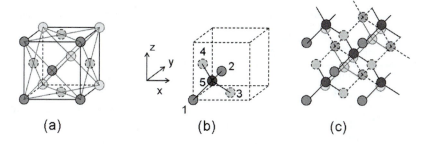

Figure 8.2. Building the crystal structure of diamond: (a) face-centered-cubic lattice, (b) diamond unit cell, (c) diamond lattice.

that yields the hardest of substances. To visualize this structure, you must first construct a cube of side a with atoms in the eight corners. You then add an extra atom in the center of each of the faces of the cube, as shown in Figure 8.2(a). Once you have visualized this, you then add an extra atom for *every* one of the atoms in the lattice. The relative location of each "new" atom is $(\frac{a}{4}, \frac{a}{4}, \frac{a}{4})$ relative to each of the "old" atoms. In Figure 8.2(b) we have drawn only four of the "old" atoms, which we label 1 to 4 in the figure. We added only one the "new" atoms, labeled 5, which is at the location $R_5 = (\frac{a}{4}, \frac{a}{4}, \frac{a}{4})$ in the coordinate frame shown. The four "old" atoms are the nearest neighbors of atom 5, with whom they share covalent bonds. Since every face-center-cubic atom has a partner, we can think of the crystal structure as two face-centered-cubic lattices of lattice constant a displaced in three dimensions by $(\frac{a}{4}, \frac{a}{4}, \frac{a}{4})$.

8.1.2 Energetics

Solid materials do not have discrete energy levels as individual atoms do. The energies are so close to each other that they blend to form "bands" of allowable energies. There are two bands: valence and conduction. Metals have electrons in the conduction band that are essentially free to move within the metal. Electrons in the valence band cannot move freely—they are tied to their host atom. An energy "gap" exists between the valence band and the conduction band. That is, an electron in the valence band needs to gain an energy larger than the energy gap to be promoted to the conduction band. In contrast to metals, semiconductors do not have electrons in the conduction band; all their electrons are tied up in the valence band. Table 8.1 shows energy gaps for several types of materials. It ranges from very small, with germanium (a semiconductor), at 0.67 eV, to very large, with carbon in the diamond form (an insulator), at 5.47 eV. Silicon (a semiconductor), ranks somewhere in between, at 1.11 eV. If a valence-band electron acquires an energy greater than or equal to the energy gap, then it can be promoted to the valence band. This acquired energy can be in the form of light, via the absorption of a photon.

The wavelength λ of a photon is inversely proportional to its energy (E) via the relationship

$$\lambda = \frac{1240 \, \text{eV} \cdot \text{nm}}{E} \tag{8.1}$$

Table 8.1 lists the photon wavelength equivalent to the gap energy of different materials. Consider diamond. To promote a valence electron in carbon-diamond, an absorbed photon must have an energy greater than 5.47 eV, or equivalently, a wavelength shorter than 226 nm. The latter wavelength is in the ultraviolet. Because carbon-diamond does not absorb any

Table 8.1.   Energy Gaps and Corresponding Wavelengths for Common Semiconductors and Diamond

Material	Energy Gap (eV)	Wavelength (nm)	Uses
Ge	0.67	1,850	Electronics
Si	1.11	1,120	Electronics, detectors
GaAs	1.43	867	Electronics, illumination, detectors
CdSe	1.73	717	Detectors
GaP	2.3	540	Illumination
CdS	2.42	512	Detection
ZnS	3.6	344	Illumination
C	5.47	227	Jewelry, cutting

wavelength from the visible part of the spectrum, 400 nm (3.1 eV) to 700 nm (1.8 eV), visible light goes straight through the material: It is transparent! In contrast, silicon and germanium have very low energy gaps, with corresponding wavelengths at 1,120 nm and 1,850 nm, respectively (both in the near infrared). As a consequence, both absorb visible photons, and can be used as visible-light detectors. Conversely, electrons in the conduction band can decay to the valence band by emitting light with wavelength equal to that gap.

8.1.3 Compounds

The elements in columns III and IV of Figure 8.1 have, respectively, three and five electrons that can be shared for covalent bonding. Elements in these two columns also form a hybrid version of the diamond lattice, called zinc-blende, in which one set of atoms forms one face-center-cubic lattice and the other type of atom forms the displaced face-centered-cubic lattice mentioned earlier. This way, an element sharing three electrons (such as Ga) and an element sharing five electrons (such as As) can form a compound (such as GaAs) with the same lattice as diamond. Similarly, with atoms in the II and VI columns, atoms in one column (for example, Zn) share their two electrons while atoms in the other column (for example, Se) share their six electrons. Table 8.1 shows popular semiconductor variations forming zinc-blende lattices with their respective energy gaps.

Other possibilities for variations in the properties of semiconductors include substituting an element from a given column with another element from the same column. For example, in the ternary alloy $Al_xGa_{1-x}As$, column III elements Al and Ga are mixed with respective fractional amounts x and $1 - x$. The electronic and optical properties of the semiconductor vary when we change x. For example, in LEDs, changing the proportion of the atoms from the same column results in a change in the band gap and, consequently, the emission wavelength. If we make substitutions of elements in both columns we can form quaternary alloys, such as $In_xGa_{1-x}As_yP_{1-y}$.

8.1.4 Doping

Semiconductors have useful applications when they are "doped." As mentioned earlier, we can make compounds with elements of one column (IV), forming a diamond structure, or elements from two columns (III and V or II and VI), forming a zinc-blende structure. Doping refers to the technique of adding controlled trace amounts of an element from an additional column. For example, a pure group IV semiconductor such as silicon can be doped with an element from group V, such as phosphor. However, phosphor has five electrons available for covalent bonding. So if a phosphor atom takes the place of a silicon atom in the diamond lattice, it will share four electrons, leaving one electron unshared. This unshared electron (often referred to as unpaired) is only loosely bound to the site, so its binding energy will be very small, of the order of a few meV.

Upon gaining this energy, the unpaired electron becomes a conduction electron, gaining the freedom to move around the material. Figure 8.3 shows this pictorially, with an Si lattice compressed to two dimensions. (We underscore that this is an oversimplification of Figure 8.2c.) The shared covalent bonds are represented by bars. The lattice on the left is of undoped silicon, and the lattice in the middle of the picture is of silicon doped with phosphorus. Another oversimplification is that we show two phosphor atoms in the same picture. Because the concentration of dopants is very low (1 in 10^7), it is rare to find them so close to each other. Doping reduces the resistance of the semiconductor tremendously. For example, at room temperature pure silicon has a resistivity of $2 \cdot 10^3$ Ω-m. When doped with tin at the level of 1 part in 10 million, its resistance goes down to 0.05 Ω-m. The dopant atoms in this case are called donors. The material doped with donors is also referred to as *n-type*.

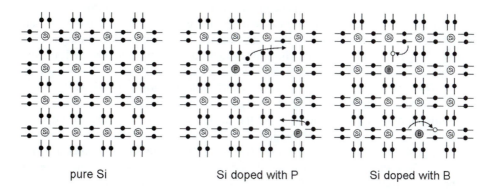

pure Si   Si doped with P   Si doped with B

Figure 8.3. Simplified view of the semiconductor lattice, showing the intrinsic silicon lattices without doping (left) and with the two types of doping: n-type (middle) and p-type (right).

The doping element can be from column III in the periodic table of Figure 8.1. In that case, the doped atom has only three electrons to share in covalent bonds with neighboring atoms. So if boron replaces a silicon atom in a silicon lattice, there is a missing electron for completing the four bonds per lattice site. This lack of an electron is called a hole. The lattice on the right of Figure 8.3 illustrates this. Electrons from neighboring bonds can hop to fill the unpaired bond, but they leave a hole behind. The net effect is that the hole can move around without much effort, involving energies also of the order of a few meV. Note that holes correspond to a lack of electrons and thus behave as positive charge carriers (although, in reality, they are not real particles). Elements that create these holes are called acceptor atoms. Semiconductors doped with acceptor atoms are called *p-type*.

When the basic semiconductor is not a column IV element (such as Si) but a column III–V compound (such as GaAs), then atoms from column VI (such as S) become donor dopants (that is, to replace an atom with five sharing electrons with one that has five sharing plus one extra). Similarly, the acceptor dopants will come from atoms in column II of the periodic table (such as Zn) sharing only two electrons instead of three.

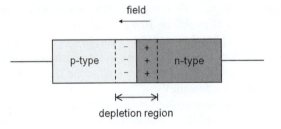

Figure 8.4. Schematic of the semiconductor diode showing the depletion layer where electrons and holes from one region diffuse to the other, establishing an electric field within the region.

8.1.5   The p–n Junction

We must reiterate that both p-type and n-type materials are electrically neutral. Their difference lies in the type of charge carriers that are involved in contributing to an electric current: holes in p-type materials and electrons in n-type materials. Upon applying an external electric field, the holes *act* like positive particles, and so they move in the same direction as the field. Conversely, electrons move against the field because they are negative.

When p-type and n-type semiconductors are put in contact, the electrons and holes from each side of the interface diffuse to the other side to satisfy the unpaired bonds and concentration. However, in doing so, they leave the regions on either side of the interface with a net charge, with the p-type side charged negative and the n-type side charged positive. This region is also known as the depletion region. Figure 8.4 shows a schematic of the junction. The separation of charges causes an electric field that grows in magnitude toward the interface, having the greatest magnitude exactly at the interface.

The diffusion of electrons and holes in the depletion region is a transient effect. It is created by the fact that electrons and holes can move around, and the concentration of either kind pushes them apart in the average. So the diffusion is pushed by the concentration of carriers. Once the electric field is developed, pointing from positive to negative, an equilibrium is established, preventing further holes from diffusing to the n-region and electrons from diffusing to the p-region.

Suppose that we apply an external field. If the external field goes from the p-side to the n-side, holes want to go to the n-side and electrons want to go to the p-side. In this case, the charge carriers want to cross the interface, so they can carry a current from one end to the other. There is still the field across the interface, but if the carriers have enough energy, they can overcome it and cross. Conversely, if we apply an external field from the n-side to the p-side, both types of carriers want to move *away* from the interface: Electrons want to pile on the n-side and holes want to pile on the p-side. Thus, the semiconductor carriers do not want to participate in a current that involves crossing the interface.

8.2   DIODES

Semiconductor junctions of the type that we have described are used to make diodes. The ideal diode is one that transmits current in one direction, with no opposition, and does not allow current to flow in the opposite direction. In essence, the ideal diode is a one-way valve. Early diodes were implemented with vacuum tubes, but the arrival of semiconductors and the p–n junction changed this dramatically, reducing the size and the electro-mechanical complexity of the device. In the last section we stated that when an external field points from the p-side to the n-side the semiconductor carriers want to cross the interface; when the field is reversed, the carriers do not want to cross the barrier. In essence, this is a diode function: When we apply a voltage that makes the p-side positive relative to the n-side, a *forward bias*, the device conducts, and when we reverse the polarity (that is, applying a *reverse bias*), the device does not conduct.

The I–V curve of an electronic component is an important design tool. It consists of the graph of the current flowing through the component as a function of the voltage applied to it. For example, the I–V curve of a resistor is the graph of $I = V/R$. It is a straight line with a slope $1/R$ that passes through the origin. The smaller the resistance, the larger the slope. Therefore, a wire with zero resistance has an I–V curve that is a vertical line. At the other extreme, a high resistance is a line with a small slope. In the limit of infinite resistance (an open circuit), the curve is a horizontal line at $I = 0$.

The ideal diode is a device that transmits all current (with no resistance) in one direction, and blocks any current in the other direction. The ideal I–V curve of a diode consists of a vertical line in the forward direction and a horizontal line in the reverse direction at $I = 0$. The left pane of Figure 8.5 shows this. The I–V curve for a semiconductor diode (right pane)

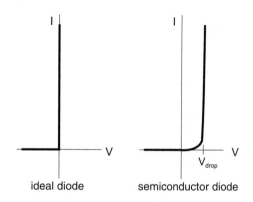

Figure 8.5. Diode I–V curves: ideal (left) and real in a semiconductor (right).

is slightly different. It conducts when forward biased. However, it is displaced due to the field across the junction. This displacement of the curve reflects a voltage drop across the diode. The current is not a vertical line like the ideal diode, but rather a curve that turns into a nearly vertical line. The current flowing through a diode for a given applied potential V_D is given by

$$I_D = I_0 \left(e^{V_D/V_T} - 1\right) \qquad (8.2)$$

where V_T is the thermal energy expressed in volts, which at room temperature is 0.026 V. The constant I_0 is called the reverse saturation current. It is a technical name that represents the carrier dynamics across the p–n barrier. For our purposes, we need to know only that it depends on the properties of the semiconductor, its doping level, and the dynamic properties of the carriers. Values of the reverse saturation current range between 10^{-15} and 10^{-13} A. The consequence of this dependence is a curve that is flat, near zero current, for low voltages. Above a certain value of the voltage, called the forward drop, the current increases rapidly. The location of the forward drop depends on the materials of the p–n junction and the temperature. Table 8.2 gives the forward drop of common p–n junctions.

In the reverse direction, the semiconductor diode conducts a finite but very small current due to diffusion (precisely, I_0), which can be ignored in most practical applications. So for most purposes, the diode biased in the reverse direction acts as an open circuit. For large

Table 8.2. Forward Drop for Common Diodes

Diode	Voltage Drop (V)
Ge	0.25
Si	0.6
GaAs	1.0
InGaN	3.0

reverse voltages, the electric potential becomes so high that at some value, a breakdown occurs and the diode conducts. This regime is not at all a failure of the device, but a new regime of operation. Electrons get liberated from the valence band, where they are spatially confined and then, once free, accelerate and gain enough energy to liberate other electrons in an avalanche fashion. In common circuits when this situation arises, the current that ensues is large enough that ohmic losses on the diode lead to its destruction. All diodes have a power rating, or a maximum power P_{max} that they can tolerate. At the reverse voltage, $P_{max} = V_{br}I_{br}$. For example, for 1N4002, the maximum power is $P_{max} = 1\,W$, which translates to about 1 A at 1 V of forward drop. It has $V_{br} = 100\,V$, so if a reverse bias has that voltage the diode reaches P_{max} with a current of 10 mA, a value that can easily be reached and exceeded. Later we describe Zener diodes, where the breakdown voltage is small and used as an operating parameter.

8.3 DESIGNING DIODE CIRCUITS

Although the physics of semiconductors is interesting, we now move into some electronics. The most accurate way to design a circuit with a nonlinear element like the semiconductor diode is using the load line method. Before we continue, we need to cover the jargon used with diodes. The general symbol for a diode is an arrow that points in the direction of current conduction (from p to n). The different variations of diodes are represented by variations of the main symbol. Figure 8.6 shows the symbols of the diodes that we discuss in this chapter.

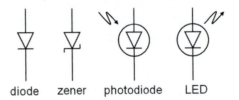

diode   zener   photodiode   LED

Figure 8.6. Symbols for different types of diodes.

8.3.1 Load Line Method

This is a method for solving an equation graphically. We use it to design circuits with nonlinear components, for which the I–V characteristics are not known analytically. The circuit of Figure 8.7 has a voltage source V_0 in series with a resistor R and a load.

Suppose that the load has an I–V curve given by

$$I = f(V) \tag{8.3}$$

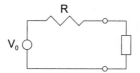

Figure 8.7. Circuit connecting a source, a resistor, and a load.

where I is the current flowing through the load and V is the voltage drop across it. Because we know the other components of the circuit, we can conclude that I and V also satisfy

$$I = \frac{V_0 - V}{R} = -\frac{V}{R} + \frac{V_0}{R} \tag{8.4}$$

The graph of Equation 8.4 is called the *load line*. We can solve for V and I analytically by combining $f(V)$ with Equation 8.4. For example, if the load were a resistor R_L, we could solve these equations easily. Using the equation for a resistor

$$I = f(V) = \frac{V}{R_L}$$

and replacing it into Equation 8.4, we get

$$V = \frac{R_L V_0}{R + R_L}$$

and

$$I = \frac{V_0}{R + R_L}$$

We can solve Equations 8.3 and 8.4 in an alternate way: *graphically*. The graphical method is a simple way of solving circuits with nonlinear elements. If we graph Equations 8.3 and 8.4, the intersection point of the two curves is the solution. Figure 8.8 shows the graphical solution for the case when $R_L = 2R$: The intersection point is $(V = \frac{2V_0}{3}, I = \frac{V_0}{3R})$.

Of course, we do not have to do all this for a resistor. However, it may be our only choice for other components, such as diodes. The most practical way to obtain the I–V curve of a diode is to graph it with an electronic instrument called the curve tracer (see Appendix C). We then draw the load line on the same graph made by the curve tracer and solve the circuit

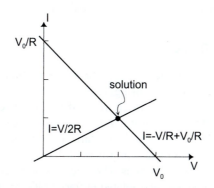

Figure 8.8. Applying the load line method when the load is a resistor.

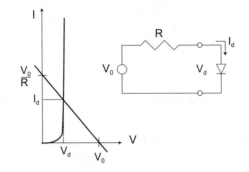

Figure 8.9. Applying the load line method when the load is a diode.

graphically. Figure 8.9 shows this. The intersection point of the diode curve and the load line, (V_d, I_d), is the solution.

In other situations, the known parameters are the supply voltage V_0 and the diode current I_d, and the unknown parameter is the resistance R. This is the case when designing circuits for LEDs. We first identify the points (V_d, I_d) and $(V_0, 0)$ on the graph. We follow by drawing a load line that passes through both points. Finally, we obtain R from the y-intercept of the load line $(0, V_0/R)$.

8.3.2 "Quick and Dirty" Circuit Design for Diodes

As shown in the diode I–V curve, when the diode is forward biased, a voltage drop V_d across it is present. If we want to make a design with a diode in series with a resistor, a less distinguished but quick design is to approximate the diode curve by a vertical line at the location of the forward drop. We then need only a rough number for V_d. The drop at the resistor is $V_0 - V_d$. Therefore, the current flowing through the circuit is

$$I \sim \frac{(V_0 - V_d)}{R} \tag{8.5}$$

8.4 DIODE FAUNA

Diodes have become a workhorse in technology. Variations of the standard diode with different geometries and components have led to a number of useful applications. This section briefly describes popular variations and applications.

8.4.1 LED and Laser Diode

Electroluminescence is probably the most significant application of semiconductors after rectification. It regards the generation of light via an electrical current. In the case of the light-emitting diode (LED), current carriers reach the diode junction and "jump down" an energy step (gap), emitting light in the process. Today LEDs cover almost the entire visible spectrum of light. They are efficient and durable light sources, gaining efficiency to such a degree that soon they will replace all forms of illumination. LEDs are diodes made with engineered semiconductor compounds. The long-wavelength (red, infrared) diodes are based on GaAs binary compounds and related tertiary (AlGaAs, GaAsP) and quaternary (AlGaInP) compounds. As we move to lower wavelengths, the n-type semiconductor shifts to P (GaP, GaAsP, AlGaP) in the green portion of the spectrum, and then to N (GaN, InGaN, AlN, AlGaN) in the blue/ultra-violet region of the spectrum. LEDs emit light when a forward current of 10–20 mA passes through them. Their forward voltage drop V_d varies between 1.5 and 4 V. For most lightly doped LEDs, the forward drop is approximately equal to the energy gap (the one the electrons jump down while emitting a photon). Therefore, $V_d \sim 1240$ V · nm/λ, where λ is the wavelength of the peak of the emission (LED emissions have a broad bandwidth).

The quick-and-dirty design described earlier is good enough for applications such as the seven-segment display that we used in the digital section. For example, if the power source is 5 V and the forward drop of the LED is 1 V, then to pass a 20-mA current through it, we need to wire the LED in series with a resistor of value given by $R = (5$ V $- 1$ V$)/20$ mA $= 200 \, \Omega$.

Another important device used widely today is the laser diode. This type of diode emits coherent and collimated laser light when a current greater than about 70 mA is supplied. Laser diodes can be used as crude light sources, as is the case of CD players or bar-code readers, or as precise light sources in applications such as spectroscopy or fundamental research. In the latter applications, stable wavelength and intensity outputs require current and temperature controls. Later in the course, we see feedback circuits that provide this type of control.

8.4.2 Photoconductor Photodiodes

When light is incident on the diode, the photons may free electrons and holes if the incoming light has enough energy (greater than the energy needed to liberate electron-hole pairs). The newly available carriers generate a current. The current does not diminish or change the junction characteristic or performance. As a consequence, the I–V curve of the photodiode is the standard diode curve, but with a vertical shift that depends on the input light (see Figure 8.10). The freed electrons and pairs find that the better path for recombination is to travel *away* from the junction, thus generating a *reverse* current. This has to do with the diffusion of carriers in the diode. As a consequence of this reverse current, the I–V curve shifts *down*. The current flowing through the photodiode can be modeled by the following equation:

$$I = I_D - I_L \tag{8.6}$$

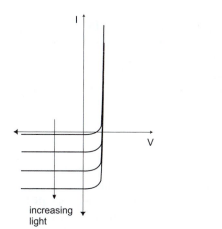

Figure 8.10. I–V curves for a photodiode receiving light.

Here, I_D is the diode curve of Equation 8.2 and I_L is the photo-current, or the current generated by the light.

Figure 8.11 shows a typical biasing circuit. Notice that the reverse bias causes the operating point to be in the third quadrant of the diode curve. This way, we can take the most advantage of the photodiode current. This point becomes important later.

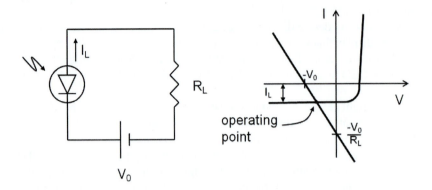

Figure 8.11. Circuit for reverse-biasing the photodiode.

The value of the load resistor depends on the type of application. In ordinary applications when we want to extract a constant voltage proportional to the light, this load can be a not-too-large resistor (for example, 1 kΩ). Otherwise, the intersection of the load line with the diode curve will be in the fourth quadrant and close to the forward drop of the diode (see the next section), where the photodiode does not behave as a pure current source. For high-speed applications, the load should be the circuit to which it is connected, which should have a 50 Ω input impedance. The cable that connects the photodiode to the load should also be a cable with a 50 Ω impedance (such as the coaxial cable RG-58). If the load device in the high-speed application has a high input impedance, such as an oscilloscope, then terminate the line with a coaxial tee and 50-Ω terminator, as shown in Figure 6.30. Later in Chapter 10 (Operational Amplifiers), we introduce a circuit that amplifies the photo-current.

Typical photodiodes have fast responses, with rise times of 0.1–10 ns and high sensitivities of about 0.1 A/W (but with maximum currents not exceeding 1 mA—check specifications). PIN photodiodes have an intrinsic (undoped) semiconductor region between the p and n regions that adds speed to the device by lowering its capacitance via the larger depletion region. A particularly modern development in photodiodes is the avalanche photodiode, or APD. These photodiodes get to be operated near or above their reverse breakdown voltage. When they detect light, a reverse breakdown occurs, generating a large current. Because of this high gain, APDs are the semiconductor version of photomultipliers, and are used for detecting single photons.

8.4.3 Photovoltaic (Solar) Cells

Solar cells are photodiodes designed to generate power. They work in the photovoltaic regime. Suppose that we have a photodiode that is connected to a high-impedance load. The resistance is near infinite, so the load line is nearly horizontal. The operating point, or the intersection of the diode curve with the horizontal axis, is near zero current. As light is incident on the photodiode, the diode curve shifts down, and the crossing point moves toward positive voltages. Thus, we have a voltage output as a function of illumination.

The Sun's luminous power is clean and readily available for us to take. On the surface of the Earth, the solar power is $1\,\text{kW/m}^2$. Solar cells are usually made of silicon and have an efficiency of 10 to 30 percent. Many aspects of the commercial solar cell (Figure 8.12) have been engineered to maximize the power generated.

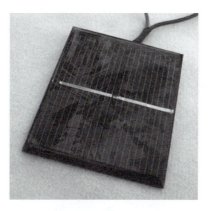

Figure 8.12. A solar cell has peculiar features. To reduce cost, the cell has multi-crystalline sections, as shown by the regions of different shades. Each region is a single crystal. It is n-type on the front side and p-type on the back. Thin wires collect charges from the front semiconductor. The other electrode is a flat metal plate on the back.

We now take a closer look at the electronics of the solar cell. In the previous two applications, with forward-biased LED and reverse-biased photodiode, the operating point of the diode was in the first and third quadrant, respectively. In both cases, the power dissipated by the diode was positive, so they took power. Consider the biasing situation shown in Figure 8.13. The photodiode (solar cell) is forward biased, but at low enough voltage that it operates in the fourth quadrant of the I–V graph. Its dissipated power $P_{diss} = -V_{sc}I_{sc}$. Because $P_{diss} < 0$, it does not dissipate power. Instead, it *generates* power.

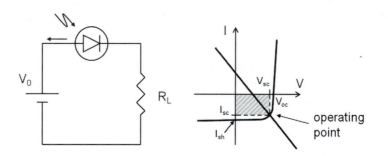

Figure 8.13. Biasing circuit for operating a solar cell.

The generated power $P_{gen} = -P_{diss}$ is the area of the shaded region in the graph. Also labeled in the graph are the crossings of the diode curve with the axes: the open-circuit voltage V_{oc} and the short-circuit current $-I_{sh}$. Depending on how the bias is done, the shaded region may not be large. For example, if V_0 is close to V_{oc}, or if the load is too high, the intersection point of the load line with the diode curve will be near the axis and the shaded region in the graph will be a thin horizontal rectangle. To maximize power, we need to maximize the shaded area, also known as the *fill factor*:

$$F = \frac{I_{sc}V_{sc}}{I_{sh}V_{oc}} \qquad (8.7)$$

A simpler solar-cell circuit does not use any forward bias. In this case, the intersection of the load line with the I–V axes is the origin. The intersection point then is purely determined by the slope of the graph, $1/R_L$, and the diode/cell curve. In a given solar application, the illumination is constant, so we can find the load that maximizes the power generated. In larger solar arrays, an impedance converter is necessary so that the solar cell sees the same load all the time.

The previous analysis is for an ideal solar cell. It is instructive to analyze the operation of a "real" solar cell. Often in electronics, one studies devices by first assuming that they are ideal, but when a serious effort is needed, the real device is modeled. Although the solar cell is a single device, it behaves as if it were the device shown in Figure 8.14. The model

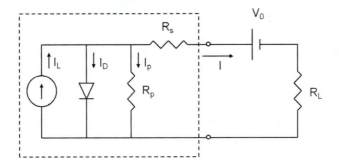

Figure 8.14.   Model of the real solar cell.

consists of a current source I_L (the photo-current), a diode, a parallel resistance R_p, and a series resistance R_s. The last two have to do with technical details of the actual device, such as the type of semiconductor, the geometry of the diode, and the amount of doping. The current I flowing out of the solar cell terminals is $I = I_L - I_D - I_p$. The voltage across the diode terminals is then $V_D - IR_s = V_0 - IR_L$.

The power that individual solar cells generate is small: If $V_{oc} \sim 0.5\,\mathrm{V}$ and $I_{sc} \sim 50\,\mathrm{mA}$, then $P_{gen} = 5\,\mathrm{mW}$. Thus, solar panels must have many solar cells. Commercial modules produce about $30\,\mathrm{mA/cm^2}$ and involve a combination of cells that are in series and in parallel. If there are N cells in series and M of these in parallel, then the operating point of the module is (NV_{oc}, MI_{sh}).

8.4.4   Zener Diode

Earlier we mentioned that when we apply a reverse-bias voltage greater than the breakdown voltage, the diode suddenly conducts. This feature can be used for practical purposes. In Zener (or breakdown) diodes, the device has been engineered to have a low breakdown voltage, and once reached, to conduct with very little resistance (that is, with a steep I–V curve), as shown in Figure 8.15. Its slope is typically a few millivolts per milliampere (or an effective resistance of a few ohms). It can be used as a rough regulator for fixed loads, or as a reference voltage. Zener diodes come in a variety of reverse-breakdown voltages, ranging

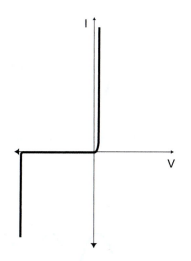

Figure 8.15. I–V curve for a Zener diode.

from as low as 1 V to as high as a few hundred volts. Later in Section 8.5.6 we design circuits that use Zener diodes.

8.4.5 More Diodes

A number of other diodes not discussed here are also important. The *Shottky* diode is the junction of a doped semiconductor (n-type) with a metal. Interestingly, such a sharp variation still behaves like a diode. The junction with the metal allows faster switching speed (0.1 to a few nanoseconds) than regular diodes (hundreds of nanoseconds) and lower forward drops (0.1–0.5 V). Shottky junctions are used to speed up transistor switching speeds in TTL logic gates. They also have their shortcomings.

When diodes are reverse biased, their capacitance depends on the applied voltage. A *varactor* diode maximizes this effect for the purpose of making frequency-tuning devices.

8.5 DIODE APPLICATIONS

Here we discuss a number of important, useful, and clever applications of diodes. Diodes once had their own time in the digital spotlight, as bearers of the diode-transistor-logic (DTL) family of digital gates and circuits. However, that technology gave way to TTL and CMOS technologies, and is now obsolete.

8.5.1 Rectification

An important application of diodes is rectification. It is used for generating a DC voltage from an AC voltage. Figure 8.16 shows the *half-wave rectifier*. The voltage across the load resistor is non-zero only when current flows through the circuit (when the AC voltage swings in the direction to positively bias the diode), as shown in Figure 8.16.

A more power-efficient arrangement is the *full-wave rectifier*, also known as a bridge (see Figure 8.17). Here the current is cleverly routed so that it flows through the resistor in the same direction for both swings of the input voltage.

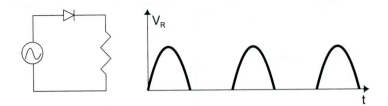

Figure 8.16. A half-wave rectifier circuit and output.

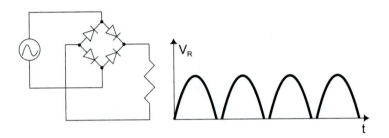

Figure 8.17. Full-wave rectifier circuit and output.

If you follow the path of the current, you conclude that the voltage across the resistor always has the same polarity. Figure 8.17 shows the voltage drop across the resistor as a function of time.

Making a DC Power Supply from an AC Source A power supply is a device that transforms power from one kind to another. Most common supplies use the utility power (110 V in North America, 220 V in Europe) to transform the power. However, these are not the only types of supplies. Some devices transform the power from other types of inputs, such as DC sources. Here we discuss how to make a DC power supply from an AC voltage via voltage rectification.

We just discussed a full-wave rectifier. The next step in making the voltage closer to a constant is to connect the output of the bridge to a capacitor and load resistor, as shown in Figure 8.18. When the voltage of the rectifier rises, it charges the capacitor. However, when

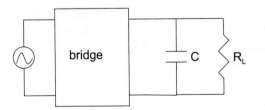

Figure 8.18. Schematic of a DC power supply made from an AC input source.

the rectifier voltage drops, the capacitor discharges through the load. It cannot discharge through the bridge because the diodes do not conduct in the reverse direction. The circuit's time constant $R_L C$ determines how fast the capacitor discharges. The voltage across the load then looks like the one in Figure 8.19: a DC voltage with a small, strange-looking ripple. We can estimate the relative size of the ripple, $\Delta V / V$, by assuming that the early part of

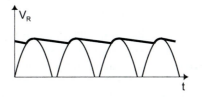

Figure 8.19. Waveform coming out of the DC power supply described in Figure 8.18.

the capacitor discharge curve is a straight line. That is, during this early part, $t \ll R_L C$, so it can be approximated by

$$V_R = V_0 e^{-t/R_L C} \simeq V_0 \left(1 - \frac{t}{R_L C} \right) \tag{8.8}$$

where we have used the binomial expansion for small values of $t/R_L C$. For a time $t = \Delta t$, the ripple can be obtained from the previous equation, yielding

$$\Delta V_R = \frac{\Delta t V_0}{R_L C} \tag{8.9}$$

The time Δt can be approximated by half of the period of the input AC: $\Delta t \simeq T/2 = 1/(2f)$. The final expression for the relative ripple voltage is

$$\frac{\Delta V_R}{V_0} \simeq \frac{1}{2f R_L C} \tag{8.10}$$

For example, suppose we want to rectify a 12 V (peak) AC voltage with a frequency of 60 Hz to a 10 percent ripple. If the load resistor is 1 kΩ, then using Equation 8.10, we find that we need a capacitor of about 100 μF.

This procedure gets us a good DC voltage only if we use large capacitors. Otherwise, the size of the ripple is determined by the load. The problem is solved by using feedback, a topic that we cover in Chapter 10 (Operational Amplifiers). The device that does this, called a *voltage regulator*, is located between the capacitor and the load, as shown in Figure 8.20. It has two requirements: (1) The input ripple not exceed 10 percent, and (2) the lowest point of the ripple must be higher than the desired final voltage. Standard regulators are labeled by the industry as 78XX, where XX specifies the regulated voltage. In the next section we give the recipe for making a DC power supply.

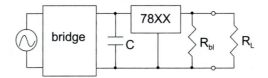

Figure 8.20. Diagram for a regulated power supply.

Designing a DC Power Supply   Suppose that we want to design a DC power supply with a final voltage V_{DC}. The latter value determines the voltage of the regulator that we choose. We must follow a few fundamental steps in the design.

1. If the input AC is the output of a transformer, choose one so that $V_{peak} > V_{DC} + 2V_{diode-drop} + \Delta V_{ripple}$. Be aware that transformers are specified in terms of their RMS voltage ($V_{RMS} = V_{peak}/\sqrt{2}$). For rectifier diodes, $V_d \sim 0.7$ V.

2. Connect a full-wave bridge to the output of the transformer.

3. The capacitor discharge time determines the ripple. A regulator tolerates a ripple of less than 10 percent. Assume a worst-case load resistor R_L (its smallest value in a given application). This gives you a maximum current. Do not go beyond the specifications of the regulator ($I_{max} \sim 100$ mA for 7805). Assuming a 10 percent ripple, you then calculate the size of the capacitor with $C = 5/(R_L f)$ (make sure you know where this formula comes from).

4. It is helpful to have a bleeder resistor (higher than the load) shunting the output of the regulator. This discharges the capacitor when the power supply is turned off.

8.5.2 Clipping

Because the diode has a forward drop V_d, it will not conduct until the applied voltage is above that drop (0.6 V, for example). If we look at the output of a circuit such as the one shown in Figure 8.21, it results in a clipping of the input when it swings above 0.6 V.

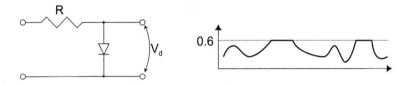

Figure 8.21. Diode clipper: circuit (left), and output waveform (right).

When considering this circuit, we need to figure out the value of the current-limiting resistor R. If the current flowing from the source is I, then

$$R = \frac{V_{in-max} - V_{cl}}{I} \tag{8.11}$$

where V_{in-max} is the maximum input voltage that we expect. We must then pick a maximum current I_d through the diode at maximum input voltage. If the load has a resistance R_L, then

$$I = I_d + \frac{V_{cl}}{R_L} \tag{8.12}$$

If for example we pick $I_d = 10$ mA and $R_L = 100$ Ω, then $I = 11$ mA. If $V_{max} = 2$ V, then $R = 127$ Ω.

We can change the clipping level by putting a voltage source V_0 in series with the diode (but negatively biasing it), as shown in Figure 8.22. This way, the input has to overcome the 0.6 V plus V_0 ($V_{cl} = V_d + V_0$).

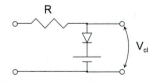

Figure 8.22. Diode clipper with a variable clipping voltage.

The circuits can also be implemented with Zener diodes, in which case, the Zener diode's breakdown voltage determines the clipping voltage. This type of clipper is even more desirable because the reverse-conducting region of the Zener is very sharp. We can also use two Zener diodes back-to-back, as shown in Figure 8.23, and provide clipping action in both polarities.

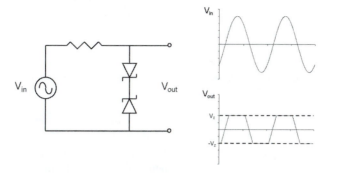

Figure 8.23. Zener clipper for both polarities.

8.5.3 Diode Clamping

We can use a clever diode circuit to modify the DC level of an AC signal. Consider the circuit of Figure 8.24. This is challenging to understand because it involves some subtleties with the capacitor and diode. Let us analyze the circuit when we first connect it to a source

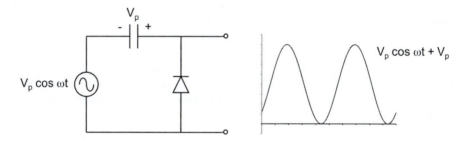

Figure 8.24. Diode clamp: circuit (left), and input and output waveforms (right).

that outputs $V_p \cos \omega t$. Let us also consider an ideal diode (without a forward drop). When the input voltage is in its positive swing, the diode cannot conduct because the diode cannot conduct in the reverse direction. However, when the input swings in the reverse direction, the diode conducts, charging the capacitor. When the input voltage reaches $-V_p$, the capacitor is charged by the maximum amount. At this point, the capacitor has its positive end toward the diode and its negative end toward the supply. When the voltage starts to decrease, the charge between the diode and the capacitor gets trapped there. As a consequence, the capacitor is left *always* charged to a potential V_p. The output voltage therefore becomes the input voltage plus the capacitor voltage: $V_p \cos \omega t + V_p$. The figure considers an ideal diode. If we take into account the real diode with forward drop V_f, then the output of the circuit is $V_p \cos \omega t + V_p - V_f$.

8.5.4 Peak Detector

Consider the circuit of Figure 8.25. It looks very similar to the previous clamping circuit. It works the same way, but the output is different because it looks at the voltage across the

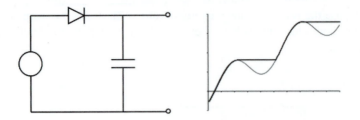

Figure 8.25. Diode peak detector: circuit (left) and output waveform (right).

capacitor. When the input voltage initially increases in the positive direction, the diode conducts and charges the capacitor. When the input voltage decreases after reaching a maximum value, the diode stops conducting because the potential of the capacitor is greater than the potential produced by the source. As a consequence, the capacitor cannot be discharged and it keeps the peak voltage of the source. If the voltage of the source swings back and rises to voltages above the previous peak, then the diode conducts and the capacitor continues to get charged until the input no longer increases. If the input potential decreases, the diode stops conducting and the capacitor ceases to be charged, retaining the previous peak potential. The circuit is then rightfully called a peak detector. This behavior is simulated in Figure 8.25.

Some technical points relate to implementing this circuit. If we connect it to a load, the capacitor will discharge, so at the very least, it has to be connected to a very high impedance load so that the discharge is not appreciable. Then you also need to be able to reset it so you can detect a "new" peak. We show how to do this later in Chapters 9 (Transistors) and 10 (Operational Amplifiers), when we discuss transistor switches and operational amplifiers, respectively.

A clever application of the peak detector is the circuit that drives the flash of a camera. It uses a circuit that interrupts an input signal from a battery, amplifies it with a transformer, and, via a peak detector, charges a capacitor. The capacitor is then discharged to give power to the flash lamp.

8.5.5 Voltage Multipliers

Diodes can be connected in even more clever ways to make voltage multipliers. Check the circuit of Figure 8.26: It is a combination of a diode clamp with a peak detector. In the negative swing of the input voltage, diode D_1 conducts and charges capacitor C_1 to V_p (assume ideal diodes for simplicity). Then in the positive swing, the input diode D_1 does not conduct, but diode D_2 conducts because C_2 is uncharged. However, C_2 sees the sum

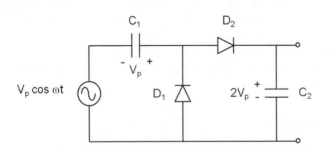

Figure 8.26. Circuit for a voltage doubler.

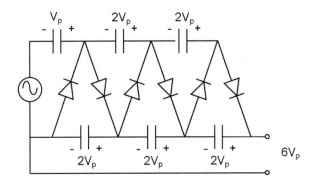

Figure 8.27.   Circuit for voltage multiplier (×6) of the Cockroft-Walton design.

of the input voltage plus the voltage of C_1, so C_2 gets charged. After a few cycles, C_2 is charged to $2V_p$ (neglecting diode drops). The circuit is called a voltage doubler. It then stays at that voltage thereafter.

The circuit of Figure 8.27 is a 3-stage Cockroft-Walton multiplier, designed by John Cockroft and Ernest Walton, who in 1932 used it to produce high voltages for particle accelerators. In general, the output for n stages is $2V_p n$ (neglecting diode drops). This circuit is used for low-current, high-voltage generators, such as oscilloscopes, air ionizers, and bug zappers.

8.5.6   Zener Regulator

You might think that we can use the voltage clipper to regulate a voltage signal that carries a ripple. Indeed, this is possible. However, this works only with an ideal diode. A real diode has an I–V curve that is not vertical. The forward drop varies a few tenths of a volt, depending on the current flowing through the circuit. Furthermore, since the I–V curve of the diode is not linear, the use of a common diode as a regulator is less predictable. The component that is closest to an ideal diode is the reverse-biased Zener diode. This diode has a predictably linear reverse-breakdown curve, shown earlier, that can be used to make a regulator.

As an example of a Zener regulator, consider the circuit shown in Figure 8.28. Assuming that we have a constant load R_L, we can estimate the resistor R and the regulating effects of the Zener.

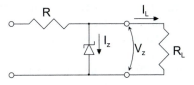

Figure 8.28.   Zener voltage regulator.

We can calculate the resistance using

$$R = \frac{V_{in} - V_Z}{I_Z + I_L} \tag{8.13}$$

The load current is $I_L = V_Z/R_L$. No prescription covers all possible cases, so the design depends on the particular application. For example, suppose we have an input voltage of 10 V with a ripple of about 1 V. We want to stabilize it to about 5 V and apply it to a 1 kΩ load. Then we can choose a Zener diode with $V_Z = 5$ V and $P_{max} = 500$ mW. The latter implies that $I_{Zmax} = 100$ mA. Assuming the Zener current is in the linear region ($I_Z = I_{Zmax}/2 \sim 50$ mA), then applying Equation 8.13 to the two extreme cases of V_{in} yields the requirement 164 Ω $< R <$ 200 Ω. To obtain the amount of regulation (final ripple), we rewrite Equation 8.13 as

$$V_Z = V_{in} - I_Z R - \frac{V_Z R}{R_L} \tag{8.14}$$

In the linear region, I_Z can be written as

$$I_Z = \frac{V_Z}{R_Z} - \frac{V_{Z0}}{R_Z} \tag{8.15}$$

where R_Z is the effective Zener resistance. Replacing Equation 8.15 into Equation 8.14 and simplifying, we get

$$V_Z = \frac{V_{in} + R V_{Z0}/R_Z}{1 + R(1/R_L + 1/R_Z)} \tag{8.16}$$

Since $R_Z \ll R_L$, then

$$V_Z \simeq \frac{V_{in} + R V_{Z0}/R_Z}{1 + R/R_Z} \tag{8.17}$$

The change ΔV_{in} as a function of ΔV_Z is

$$\Delta V_Z = \frac{\Delta V_{in}}{1 + R/R_Z} \tag{8.18}$$

For our example, then, $\Delta V_Z = \Delta V_{in}/100$ for $R_Z = 2$ Ω. Therefore, the Zener regulator improves the regulation from 10 percent to 0.1 percent. If the load varies, we must design R to be within the worst-case situations.

Zener diodes are not normally used as regulators because feedback-based voltage regulators can do a much better job. Zener diodes are more often used as voltage references driven by current sources, an application that we will see in Chapter 10 (Operational Amplifiers).

8.5.7 Touch Sensors

Diodes in combination with capacitors make a very interesting circuit: a switch that responds to touch. You see these types of switches often, especially in elevators, laptop computers, iPods, and modern cellphones. Don't you ever wonder why in those switches you do not really "push" anything? Indeed, the mechanism is neither optical nor mechanical: It is electrical.

Consider the circuit of Figure 8.29. A square digital waveform is input to a NOR gate. One of the inputs of the NOR gate is the touch sensor. After the NOR gate, we have a half-wave rectifier, which feeds to a CMOS NOT gate. We use CMOS gates because they have very high input impedance.

For now, we ignore the circuit associated with the NOR gate. If the input waveform goes through the NOR gate, then the capacitor C of the half-wave rectifier is periodically charged. Note that the role of the diode is the same as that of a half-wave rectifier: to only charge the capacitor. If the input charging frequency f is high enough (with a period that

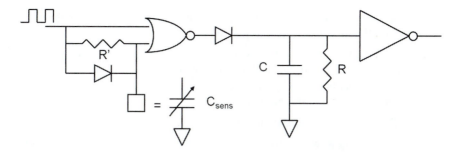

Figure 8.29. Circuit for a touch sensor.

is shorter than RC), then the capacitor does not have enough time to discharge through the resistor R to a voltage below the low input threshold of the gate (say, 3.5 V). Then the input to the NOT gate is always 1, and so its output is 0.

If the capacitor does not get charged fast enough, it has time to discharge and lower its potential to the one that corresponds to a logic zero. If it does not get any charging pulses, it quickly becomes fully discharged. In this case, the output of the NOT inverter becomes a logic 1.

Now we analyze the circuit connected to the NOR gate. (Just as a reminder, the output of a NOR gate is 1 only when both inputs are 0.) An input signal of frequency f sends two inputs to a NOR gate. One input is directly connected to the NOR gate. The other branch of the connection connects to the input of the NOR gate via a resistor R' in parallel with a forward-biased diode. The touch-pad input to the circuit is connected to the second input of the NOR gate. If we ignore the touch pad, when the input is high, the second input of the NOR gate gets charged through the diode; when the input goes low, the NOR input discharges through R'. Therefore, both inputs are 0.

The touch pad acts like a capacitor of value C_{sens}. The pad is a metal plate with a thin insulator over it. When nothing touches the pad, its capacitance to ground is very small. Thus, when the capacitor is charged, it discharges quickly through R'. When this happens, the two inputs of the first NOR gate are the same and the gate acts like an inverter, sending the input signal to the next stage. If we touch the pad, we increase the capacitance to ground and, therefore, increase the discharge time constant of the touch-pad capacitor. When the input waveform switches to zero, the longer discharge time of the touch-pad capacitor keeps the second input of the NOR gate high. This makes the output of the gate low, allowing time for the rectifier capacitor C to discharge through the resistor R. Thus, when we touch the pad, the output of the circuit is 1. When we do not touch the pad, the output is a 0: a touch switch.

8.6 PROBLEMS

1. Consider the diamond structure of Figure 8.2. The coordinates of the nearest neighbors to atom 5 in pane (b) of the previous figure are $R_1 = (0,0,0)$, $R_2 = (\frac{a}{2},0,\frac{a}{2})$, $R_3 = (\frac{a}{2},\frac{a}{2},0)$, and $R_4 = (0,\frac{a}{2},\frac{a}{2})$. Find the coordinates of atoms 1, 2, 3, and 4 relative to atom 5, and show that the four distances are the same.

2. Consider the p–n junction of Figure 8.4.

    (a) What is the direction of the electric field in the p–n junction? Redraw the figure with the electric field.

(b) Since there is an electric field, there is an electric potential across the semiconductor. Which side is at higher potential?

(c) Assume that, on either side of the interface, the charge density is uniform from the interface to the end of the depletion region, positive on the n-side, and negative on the p-side. Is the potential a linear function of the distance along the axis of the device? Justify your answer.

3. An LED has the I–V curve shown in Figure 8.30.

(a) If this LED lights well at 10 mA, use the load line method to determine the biasing circuit using a 4-V battery.

(b) Repeat the design using the "quick-and-dirty" method.

(c) If we use a forward bias of 3 V and a 500-Ω resistor, find the current flowing through the diode.

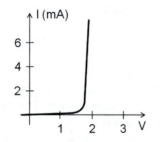

Figure 8.30.   Diode curve for Problem 3.

4. A 15-V DC power supply is connected to a diode in series with a 750 Ω resistor. The I–V curve of the diode is shown in Figure 8.31. If the diode is forward biased, answer the following questions:

(a) How much current is flowing through the diode?

(b) What is the voltage drop at the diode?

(c) How much power is dissipated by the resistor?

(d) How much power is dissipated by the diode?

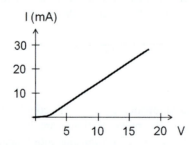

Figure 8.31.   Diode curve for Problem 4.

5. A 10-W source (AC) has a peak voltage $V_p = 10$ V. The source is connected to a full-wave bridge with diodes that have a forward drop $V_d = 0.7$ V and a power rating

$P_{max} = 1$ W. The load resistor is 1 kΩ. The output waveform is similar to the one of Figure 8.17.

(a) Explain the small flat regions of zero voltage in the waveform of the figure.

(b) What is the maximum voltage across the load resistor?

(c) If one (either) end of the AC supply is connected to one (either) end of the resistor, some part of the circuit will go up in smoke (literally!). Identify the part and explain why.

6. In the circuit of Figure 8.32, the input is a 60-Hz voltage (AC) with an RMS value of 8.5 V. The bridge has diodes with a forward drop of 0.7 V. The capacitor has a value of 20 μF and is connected to a load resistor shown.

(a) What is the minimum value of the load resistor R_L so that the ripple voltage does not exceed 10 percent of the peak output voltage?

(b) Make a careful sketch of the voltage across the load when it is 600 Ω. Label your axes.

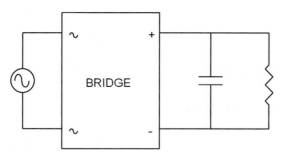

Figure 8.32.   Circuit for Problem 6.

7. Design a +12 V power supply:

(a) What should be the RMS voltage of the output of the transformer?

(b) If the maximum output current is 0.5 A, what is the value of the capacitor?

(c) Make a full diagram of the power supply (include a regulator).

8. Consider the circuit shown in Figure 8.33. The input voltage is sinusoidal (AC), with a peak amplitude of 3 V. The resistor has a value of 1 kΩ. When conducting, the two diodes have a forward drop of 0.7 V. Make a sketch of the output voltage. Carefully label your axes.

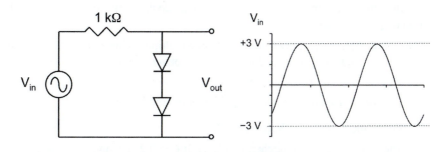

Figure 8.33.   Circuit for Problem 8.

9. Design a clipping circuit that limits the voltage of an input signal to within the range $-5\,$V to $+5\,$V.

10. Consider the clamping circuits of Figure 8.34. Assume a steady state and neglect the diode drops.

   (a) What would be the output waveform in the circuit (a)?

   (b) What would be the output waveform in the circuit (b)?

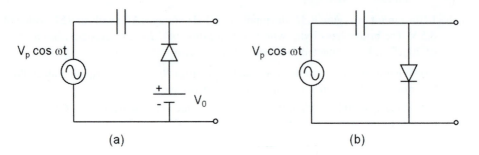

(a)   (b)

Figure 8.34.   Circuits for Problem 10.

11. In the circuit of Figure 8.35, the input voltage is sinusoidal and has a peak voltage of $5\,$V. The diode has a drop $V_d = 0.7\,$V.

   (a) Make a sketch of V_{out} as a function of time, neglecting V_d. Carefully label your axes.

   (b) Make a sketch of V_{out} as a function of time, accounting for V_d. Carefully label your axes.

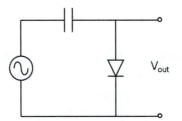

Figure 8.35.   Circuits for Problem 11.

12. In the circuit of Figure 8.36, the power source has a peak voltage of $10\,$V. Find the output voltage. Assume for simplicity that the diode drops are zero. Find V_{out}. Make a sketch, if necessary.

13. The circuit of Figure 8.37 is a Cockroft-Walton multiplier. The input signal is a square wave with a high voltage V_p and a low voltage of *zero* volts.

   (a) What is the output voltage? (Neglect diode drops.)

   (b) If $V_p = 5\,$V, find the output voltage, including diode drops of $0.6\,$V.

14. Consider a touch sensor of Section 8.5.7.

   (a) The pad is untouched (neglecting C_{sens}). If $C = 0.22\,\mu$F and $R = 10\,$kΩ, what is the minimum value of the input frequency for the output to be always low?

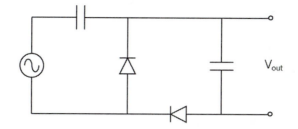

Figure 8.36.   Circuits for Problem 12.

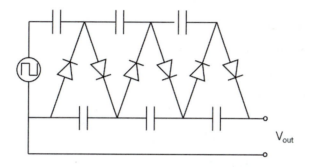

Figure 8.37.   Circuit for Problem 13.

(b) If $R' = 512\,\text{k}\Omega$, what is the minimum value of C_{sens} so that, with the frequency of the previous part, the output of the first NOR gate is always low?

(c) Make a sketch of the waveforms of the inputs and outputs of both NOR gates for two cases: when the pad is untouched and when the pad is touched.

15. Consider the circuit of Figure 8.38. The OR gate accepts any voltage greater than or equal to 2.5 V as a logic 1 and anything below 2.5 V as a logic 0. Its logic 1 output is 5 V and logic 0 output is 0 V. The input voltage is a TTL square wave with a frequency of 1 kHz. The resistor is 1 kΩ, and the capacitor is 0.5 μF. Neglect the forward drop across the diode.

(a) Make a sketch of the voltages of points A, B, and C as a function of time.

(b) Find the range of frequencies for which the output of the gate is a logic 1 *permanently*.

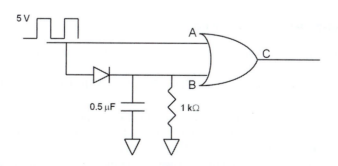

Figure 8.38.   Circuit for Problem 15.

8.7 LAB PROJECTS

This section presents ideas for lab projects.

8.7.1 I–V Curve

Required equipment: Curve tracer (if a commercial one is not available see Appendix C for a home-made one).
Required components: Silicon diode (1N4005); red, green and blue LEDs; 5-V Zener diode.
Measure the I-V curve of:

- Standard (IN4002) diode (forward bias)
- Red LED (forward bias)
- Green LED (forward bias)
- Blue LED (forward bias)
- 5-V Zener diode (reverse bias)

Recognizing the ends of a diode:

- A bar on one end of tubular diodes marks the cathode.
- The longer leg of an LED denotes the anode.

8.7.2 Diode Clamp

Required equipment: Function generator, oscilloscope.
Required components: Silicon diode (1N4005), resistor.
Wire the circuit shown. Look at the output waveform with the scope and explain it fully.

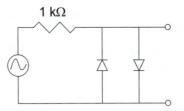

8.7.3 Make-and-Take LED Flasher

Required equipment: Soldering equipment: iron, solder wire, shrink tubing, heat gun. The work area must be well ventilated.
Required components: Bright-color LED, CR 2016 3-V battery, battery holder, small push-button switch.
The objective of this section is to learn to solder. In reward, each student can take his or her trophy home. LEDs powered by a 5-V supply need a resistor to drop the voltage on the diode to a value that corresponds to the expected current. For the LED, this current is between 10 mA and 20 mA. Although the resistor drops the voltage to what is needed for the diode, it also consumes power. If we use a battery, we want to use all of the battery power

on the LED. In this project we will use a 3-V battery so that we can connect it directly to the LED.

The components to be assembled are a mercury battery, a battery holder, an LED, and a push-button switch. Figure 8.39 shows these, together with the finished product.

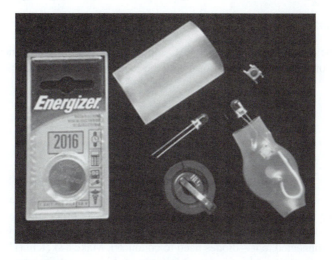

Figure 8.39.   Components for the make-and-take LED flasher.

1. First assemble a circuit on a breadboard. It should have the battery in its battery holder. Using the multimeter, measure the current flowing through the diode and the voltage drop across the diode. Measure the voltage across the battery when no current is flowing through the circuit. Is there any difference? Why?

2. Presoldering: Bend all metal contacts so that they can be hooked to each other.

    (a) Take the battery off the holder.

    (b) With the pliers supplied, bend the legs of the battery holder and switch.

    (c) Place the LED next to the battery holder and estimate the places where you should bend the legs of the LED in the shape of a hook so that the longer leg hooks to the anode of the battery holder and the other leg hooks to the switch.

3. Plug in the soldering iron, and when it is hot, solder the parts in the following order: (1) LED to the anode terminal of the battery holder, (2) LED to the switch, and (3) switch to the negative terminal of the battery holder. The best technique involves first putting some solder on the heated iron and heating the two metals to be joined with the iron. Optimum soldering temperature (about 200°C) is reached when the solder looks wet. Then place more solder on the place where both metals hook with each other. Do not use too much. The final weld should look shiny. *Warning*: do not breathe the fumes; they are toxic! Fans should be set up to circulate the air.

4. We will also provide transparent shrink tubing for those who want to provide a cover for their LED flasher.

8.7.4   Application: A Regulated Power Supply

Required equipment: AC source (6.3 VAC vacuum tube filament supply or 12 V transformer), oscilloscope.

Required components: Full-wave bridge, 7805 regulator, capacitors, resistors.

The most important application of diodes is rectification. It is also the main ingredient of a power supply. A regulated power supply consists of a transformer, a bridge rectifier, a capacitor, a bleeder resistor, and a regulator.

The Unregulated Supply

1. Build the circuit shown. The 6.3 VAC supply is *not* the function generator. For the purpose of observing the waveform with the scope, connect a 10 kΩ resistor as a load.

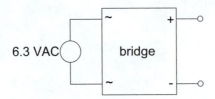

The full-wave bridge is supplied as an integrated package. Measure the voltage output of the 6.3 VAC supply with the scope. What is the peak output voltage of the supply? Measure the output of the bridge and explain its shape (make sure that you have a load resistor connected to it—diodes need to conduct!). Make a sketch. (Use the scope in the *DC* setting! Should we use the *LINE* trigger setting of the scope?)

2. Calculate the value of the capacitor for the following requirements: a maximum ripple voltage of 1V and a load of 830 Ω. The latter resistance is the design worst-case load (you will not use a 830 Ω resistor). Connect the capacitor to the output of the bridge and use the 10 kΩ resistor as the load. Sketch the waveform that you observe on the scope. Measure the ripple voltage and compare it to what you expect. Repeat the procedure for loads of 1 kΩ and 500 Ω. Explain the drawbacks of the unregulated supply.

The Regulator   Connect the 7805 voltage regulator to the output of the previous circuit, also introducing a 10 kΩ bleeder resistor (see Figure 8.40). Measure the ripple voltage for the loads applied to the unregulated supply. Compare the regulated and unregulated supplies and state your conclusions.

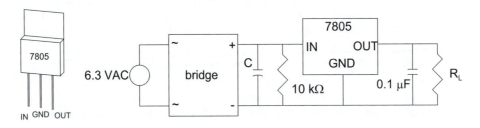

Figure 8.40.   Regulated power supply.

8.7.5   Zener Diode Circuits

Required equipment: Function generator, oscilloscope.
Required components: 5-V Zener diode, resistor.

Construct the circuit using the function generator as the power source and a 5-V Zener diode.

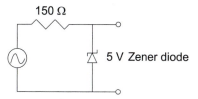

150 Ω

5 V Zener diode

Connect channel 1 of the scope to the input and channel 2 to the output. Set the amplitude of the function generator to a triangular waveform with $0.5\,V_{p-p}$. When you see both waveforms, slowly increase the DC offset of the output to $+7\,V$. How does the output compare to the input? Explain the shape of the "regulated" output.

8.7.6   Solar Cells

Required equipment: Voltmeter, ammeter, high-power lamp.
Required components: Solar cell, 0–100 Ω variable resistor.

1. Illuminate the solar cell and measure the I–V curve in the fourth quadrant. Measure V_{oc} and I_{sc}.

2. Connect the solar cell through an ammeter to the resistor. The power delivered by the solar cell is given by $P = I_{sc}V_{sc}$. Use different values of R_L to measure the delivered power. Find the optimal operating value of R_L so that the power delivered is maximum.

8.7.7   Practicum Test

Set the function generator to 10-V peak.

1. Rectify the waveform to a 10 percent ripple for a load of 100 Ω.

2. Use the rectified output to drive an LED. The current through the LED should be about 20 mA but not greater than 25 mA.

CHAPTER 9

TRANSISTORS

Contents

Next in our study of electronics is a device that rocked the world: the transistor. It made possible the miniaturization of bulky, vacuum-tube electronics technology. Today virtually every electronic device has a transistor operating within, either individually or in large groups as part of integrated circuits. Thus, transistors are the basis of modern electronics. They are at the core of amplification, switching, and regulation. Transistors can be divided into two major types: Bipolar-Junction (BJT) and Field-Effect (FET). The latter has a more modern incarnation in the metal oxide semiconductor, also known as the MOSFET. The BJT has been king for decades, but technology has made the MOSFET a rightful successor. Although it is widely accepted that BJTs are on their way out, they are still a fundamental electronic component that does not need to be replaced in many applications. In view of this, we will pay homage to the old geezer, but keep in mind that modern circuits should use the younger hot shot.

BJTs and FETs are constructed differently and also operate differently, but they provide the same function: controlling the flow of current. They have two electrodes, an input and an output, through which current can flow. A third input controls the amount of current. BJTs and FETs differ in how they effect this control: BJTs use a current, whereas FETs use a voltage; the circuitry associated with driving them is different; and BJTs consume more power. Because of the power consumption, modern circuits are based on FETs. FETs have evolved a lot since they were first invented, but their accomplishments came after years of design with BJTs. For this reason, BJTs are still the basis of conventional electronic circuits.

We start our discussion the traditional way, by first covering BJTs and following this discussion with the one on FETs. In this chapter, we discuss how the transistors work and how to use them to design simple circuits.

9.1   THE BIPOLAR-JUNCTION TRANSISTOR

To understand how a transistor works, we need to review how a photodiode works. It consists of two semiconductors, n-type and p-type, joined together. After the device is fabricated, charges rearrange, and an electric field develops across the p–n junction. Electrons and holes in this field region want to move away from the junction, but in opposite directions. A reverse bias only magnifies this depletion region, making it wider. When a photon of energy greater than the semiconductor gap energy is incident in the region and absorbed, an electron is promoted from being tied to an atom (in the valence band) to being free (in the conduction band). Because the promoted electron leaves behind a vacancy (a hole), photons effectively create electron and hole pairs. This is illustrated in Figure 9.1. The field carries these electrons and holes, creating an electrical current. An important point is that this photo–current depends only on the amount of light. The bias merely establishes the field and carries the light-induced electrons and holes to form a current. In the depletion region, the field is strong enough to prevent the recombination of the electron and the hole.

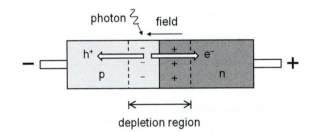

Figure 9.1.   Schematic of a reverse-biased p–n junction when a photon creates an electron-hole pair.

A transistor works the same way, only that the current carriers are injected electrically instead of luminously like in the case of the photodiode. Figure 9.2 shows a conceptual diagram of the transistor. The middle section (p-type) is the base (B) and the n-type sections are the emitter (E) and the collector (C). The p–n junction connecting the base and the collector of the transistor is reverse biased. The base is deliberately much thinner than the other two other regions. Figure 9.2 also depicts the base-collector depletion region.

Any charge carriers that land in this region are swept by the field in the region and generate a current. The emitter-base junction is forward biased, so electrons in the emitter want to go to the base.[1] As electrons enter the base and move into the base-collector depletion region, they are swept away into the collector. In fact, most of the electrons injected into the base are swept into the collector. This is known as the *transistor action*.

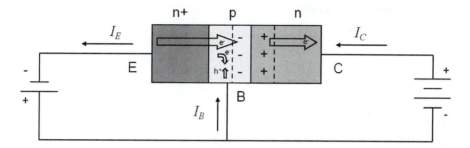

Figure 9.2. Schematic of an npn transistor.

More is involved, however. One key ingredient of the transistor action is the ratio of the base to collector currents. Because the base is p-type, it has holes as carriers. If there were no collector, all these holes would combine with the emitter electrons and form an ordinary forward-biased (base-emitter) current. However, because electrons and holes are moving in random directions with a predominant drift in the direction specified by the field, they take a certain time to recombine and actually form the emitter-base current.[2] In the transistor, this time is about 100 times longer than the time it takes for an electron entering the base to be swept into the collector. As a consequence, 1 of every 100 electrons from the emitter ends up participating in the emitter-base current. Figure 9.2 illustrates this. Transistor action is the generation of a large emitter-collector current with a small base current.

The biasing configuration of Figure 9.2 is also called the *common-base* circuit. Early transistors were made of a chunk of semiconductor (the base), with emitter and collector wires stuck to the base. We have not mentioned other details: The emitter is doped more heavily than the collector so that more electrons are available, the collector near the electrode is doped heavier than elsewhere to increase speed, and the collector and base are doped lightly to keep the reverse breakdown voltage high—the junction is reverse biased after all. In practice, the transistor looks like anything but the sandwich of Figure 9.2. Figure 9.3 shows a couple of simplified designs. The one on the left is common for individual transistors, and the one on the right is optimized for use in an integrated circuit. These are, of course, overly simplified.

Finally, since we are explaining what the carriers are actually doing, we briefly broke our rule of not following the current carriers. Therefore, now that we understand the operation of the transistor, we should go back to thinking in terms of currents, and ignore the actual carriers.

---

[1]If this sounds confusing, it is because electrons go in the opposite direction of the positively defined current. Another way to look at this is that the negative terminal of the supply on the left is connected to the emitter, so electrons want to go into the emitter.

[2]A p-to-n current requires that electrons from the n-region combine with holes from the p-region.

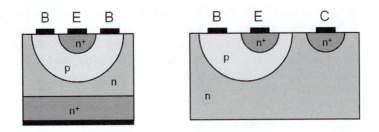

Figure 9.3. Two layouts of real transistors: when built individually (left), and within an integrated circuit (right).

9.1.1 Operation of the BJT

From the previous section, we know that transistors consist of three semiconductor regions. The outer layers, emitter and collector, have distinct sizes and amounts of doping. We concentrated on npn, but there is also pnp. The directions of currents, types of carriers, and biases are reversed when going from npn to pnp. Figure 9.4 shows their symbols.

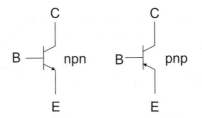

Figure 9.4. Electronic symbols for the two types of BJT transistors.

Transistor action has two requirements:

1. The base-emitter junction is forward biased so that a current I_B is generated. Once this junction is conducting, $V_{BE} \sim 0.6$ V (a diode voltage drop).

2. The base-collector junction is reverse biased.

Transistor action then translates into the relationship

$$I_C = \beta I_B \tag{9.1}$$

where β is the current gain, which is typically 100. For completeness, the emitter current is then given by

$$I_E = I_B + I_C = (1 + \beta)I_B \tag{9.2}$$

Figure 9.5 graphs a set of characteristic curves of the transistor. Those graphs show the collector current I_C as a function of the collector-emitter voltage V_{CE}, for different values of the base current I_B. When the transistor–action conditions are not met, the transistor is off and $I_C = 0$. This situation is called cutoff. Now assume that V_{BE} is forward biased and I_B is above some minimum value. As the collector-base junction becomes reverse biased, which we represent here by increasing V_{CE}, transistor action begins to unfold: I_C rises sharply to some value after which it becomes flat—it still rises, but much more slowly. Although this

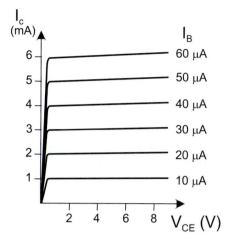

Figure 9.5. Typical transistor curves for an npn transistor.

sounds backward, the fast rise of the curve is called saturation. The flat region is said to be when the transistor is operating normally.

Consider a particular situation as an example of normal operation: $I_B = 30\,\mu A$ and $V_{CE} = 4$ V. In this case, we can read the collector current off the graph: It is about 3 mA. Note that $\frac{I_C}{I_B} = 100$. Because the curve is neither a perfect straight line nor constant for increasing values of V_{CE}, the relationship between I_C and I_B (Equation 9.1) is only approximate. As I_B is increased further, the flat section of the transistor curve rises more rapidly as well.

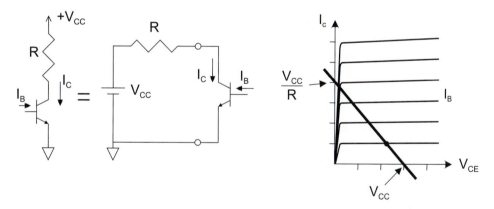

Figure 9.6. Transistor circuit (left) and its analysis using the load line method.

If we bias the collector-emitter, we can use the load line to obtain the operating parameters. Figure 9.6 shows a biasing example. For a given base current, the operating point is the intersection of the load line with the transistor curve. For small base currents (a few microamperes or less), the transistor is basically off (cutoff region): $I_C = 0$ and $V_{CE} = V_0$. Conversely, if I_B increases above a certain value, the transistor is in the saturation region, where it acts as a short circuit: $V_{CE} \sim 0$ and $I_C \sim I_{Cmax}$, where

$$I_{Cmax} = \frac{V_{CC}}{R}$$

(9.3)

9.1.2 The Transistor Switch

Following up on the last section, if for a given bias circuit we send a large base current, then we find that $V_{CE} \sim 0$, making the transistor act like a switch. So the criteria for it being a switch is

$$I_B > \frac{I_{Cmax}}{\beta} \tag{9.4}$$

When this is true, the relationship $I_C = \beta I_B$ does *not* hold because the load line intersects the rising part of the transistor curve. In this saturation mode, $V_{CE} \sim 0.2$ V. Figure 9.7 shows a sample circuit. To calculate the base current, we use

$$I_B = \frac{V_{CC} - V_B}{R} \tag{9.5}$$

and assume that $V_B = 0.6$ V (a diode drop above ground).

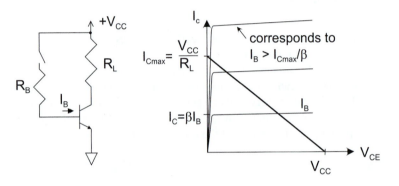

Figure 9.7.   Operation of the transistor switch.

■ **EXERCISE 9.1**

Use the transistor curves of Figure 9.5 to design a transistor switch when $V_{cc} = 3$ V and $R_L = 1000\ \Omega$. Find R_B.

The only inconvenient aspect of this circuit is that the load is connected to the supply voltage on one end and to the collector of the transistor on the other end; the load is pulled to ground only when the transistor is on. This is useful only for loads that do not need to be tied to ground permanently. Otherwise, the transistor switch can be used for any load. Even if the load is low, power transistors can deliver all the current that we need. We need to know only two parameters: the load resistance and the supply voltage.

9.1.3 The Emitter Follower

Consider the circuit of Figure 9.8. It works in a unique way. If V_{in} is greater than 0.6 V, the transistor is on and the base current, which we do *not* know, generates an emitter current I_E so that the drop across R is

$$V_{out} = V_{in} - V_{BE} \tag{9.6}$$

with $V_{BE} = 0.6$ V.

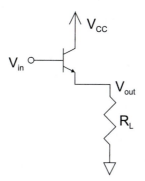

Figure 9.8.   Circuit for the transistor emitter-follower.

■ **EXERCISE 9.2**

Suppose that we apply an input voltage $V_{in} = 2\cos 2\pi ft$, with $f = 1\,\text{kHz}$, $V_{cc} = 12$ V, and $R_L = 10\,\Omega$. Make a sketch of V_{out} as a function of time. Label your axes accurately.

At first sight, the circuit of Figure 9.8 looks rather silly, but after careful analysis, we can see its great usefulness: as an impedance transformer. To appreciate the impedance-transforming property of the circuit, consider the *changes* in the signals. Suppose that we want to connect R_L as a load to V_{in} through the transistor in the emitter-follower configuration. The change in the emitter current is

$$\Delta I_E = \Delta\left(\frac{V_B - 0.6\text{V}}{R_L}\right) = \frac{\Delta V_B}{R_L} \tag{9.7}$$

But ΔI_E is also given by

$$\Delta I_E \simeq \beta\Delta I_B. \tag{9.8}$$

Combining Equations 9.7 and 9.8, we get

$$\frac{\Delta V_B}{\Delta I_B} \simeq \beta R_L \tag{9.9}$$

The input impedance of the transistor is the impedance seen by the base, as shown in Figure 9.9. Because the input impedance of the circuit is

$$r_{in} = \frac{\Delta V_B}{\Delta I_B} = \beta R_L \tag{9.10}$$

the circuit enhances the resistance of the load by a factor β.

■ **EXERCISE 9.3**

For the previous exercise, find the impedance seen by the source.

Conversely, suppose that the source has an output impedance R_S, as shown in Figure 9.10. Solving for V_E, looking toward the source we get

$$V_E = V_{in} - I_B R_S - 0.6\text{V} \tag{9.11}$$

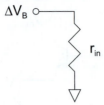

Figure 9.9.   The base sees the input impedance of the transistor, which enhances the impedance of the load.

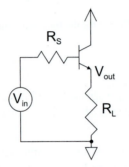

Figure 9.10.   A source with internal resistance R_S driving the transistor follower.

or

$$V_E = V_{in} - \frac{I_E R_S}{\beta} - 0.6V \tag{9.12}$$

For changes in the signals, Equation 9.12 becomes

$$\Delta V_E = \Delta V_{in} - \frac{\Delta I_E R_S}{\beta} \tag{9.13}$$

The previous equation corresponds to a source in series with an output impedance, as shown in Figure 9.11, or, equivalently,

$$r_{out} = \frac{\Delta V_E - \Delta V_{in}}{\Delta I_E} = \frac{R_S}{\beta} \tag{9.14}$$

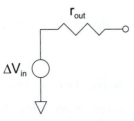

Figure 9.11.   The load sees the output impedance of the source-follower, which reduces the internal resistance of the load.

■ **EXERCISE 9.4**

If $R_S = 50\ \Omega$, what source impedance will the load see?

As we just saw, the emitter-follower's impedance-matching properties work for both the input and the output: The input sees a larger load, and the output sees a smaller source impedance. These relations can be generalized for complex impedances as well.

Finally, you might feel that we are overstating the claims of usefulness of this scheme because it has some apparent limitations. For example, V_{in} must always be above +0.6 V. Many times, we want to connect the output of an audio signal with high output impedance to an 8 Ω speaker. The follower transforms the impedances, but it would clip all the negative swings of the source, which is, of course, unacceptable. The pnp transistor works oppositely to the npn transistor: The base current comes out of the base (when $V_{BE} = -0.6$ V), the collector is more negative than the base and emitter, and current goes into the emitter and out of the collector. Therefore, one solution is to put an npn and a pnp transistor back to back in the "push-pull" configuration, shown in Figure 9.12. This way, when V_{in} is positive (above 0.6 V), the npn transistor is on and the pnp transistor is off; conversely, when V_{in} is negative (below −0.6 V), the pnp transistor is on and the npn transistor is off. This is only partially acceptable because the push-pull does not conduct for signals between −0.6 V and +0.6 V. In the next chapter we cover how to use the push-pull without this problem.

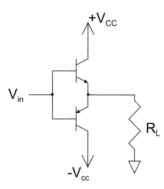

Figure 9.12.   Transistors working in the push-pull configuration.

■ **EXERCISE 9.5**

Suppose that we apply an input voltage $V_{in} = 2\cos 2\pi ft$ to the previous circuit, with $f = 1\ \text{kHz}$, $V_{cc} = 12$ V, and $R_L = 10\ \Omega$. Make a sketch of V_{out} as a function of time. Label your axes accurately.

An alternative to using the push-pull is to use a circuit that puts the signal through the transistor. One requirement is that the input be purely AC by design. The method is called biasing the transistor, which we mention again in Section 9.1.6. Figure 9.13 shows the scheme. It first uses a capacitor C_1 to remove any DC component. The cutoff for this filter is

$$\omega_1 = \frac{1}{R_{12}C_1} \tag{9.15}$$

where $R_{12} = \frac{R_1 R_2}{(R_1 + R_2)}$ is the equivalent resistance of R_1 and R_2 in parallel. The latter two resistors are designed to raise the DC level of the input AC. This way, the base has a voltage

$$V_B = V_{in} + \frac{R_2}{R_1 + R_2} V_{CC} \tag{9.16}$$

When the signal is at the emitter, we remove the DC component with C_2 acting as a high-pass filter with R_L. The design requires a few important considerations: deciding on R_1 and R_2 to add a DC offset to the input signal so that it fully passes through the transistor, and selecting R_E so that the output is not clipped. We could devote extensive coverage to this, but for most purposes, we can avoid this complicated design with operational amplifiers, which we cover in the next chapter.

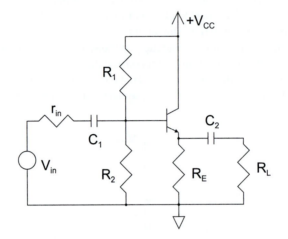

Figure 9.13.   Biasing for the emitter-follower.

9.1.4 Current Source

So far, we have been discussing voltage sources. In many applications, having current sources is desirable. These sources put out whatever voltage is necessary to keep a certain current constant. With transistors, we can make a current source that puts out a current proportional to an input voltage. This type of device is also called a *transconductance amplifier*. (A device that produces a voltage from a current is a *transresistance amplifier*, also known as a resistor.) Because the gain has units of inverse resistance, it is called siemens or mho (ohm spelled backwards). The circuit shown in Figure 9.14 is a current source. It is remarkably simple: V_B sets V_E via $V_E = V_B - 0.6$ V, which, in turn, forces a current I_C to flow through the load so that $I_C \sim I_E = \frac{V_E}{R_E}$ or

$$I_C \simeq \frac{V_B - 0.6\text{V}}{R_E} \tag{9.17}$$

So V_B and R_E decide how much current flows through the load. There is a catch though: It will not work for all loads because

$$I_C R_L + V_{CE} + V_E = V_{CC} \tag{9.18}$$

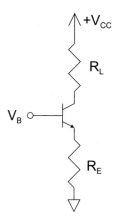

Figure 9.14. Transistor current source.

For example, suppose we have a 12-V supply powering a current source with $R_E = 500\,\Omega$. If the load is $100\,\Omega$, then the sum of the drops at the two resistors cannot exceed 12 V. Therefore, the maximum current is $\frac{(12\text{V})}{(600\Omega)} = 20\,\text{mA}$. This assumes that $V_{CE} = 0$, which is close to correct at full saturation. More accurate is to assume that $V_{CE} = 0.2$ V at maximum current. The current through the load will be $\frac{(V_B - 0.6\text{V})}{(500\Omega)}$, provided that V_B does not exceed 10.4 V.

9.1.5 The Voltage Amplifier

We started our discussion of transistors by focusing on the base current controlling the collector current via Equation 9.1. However, β is not a reliable parameter. We got around this by specifying the base voltage, which specified the emitter voltage and controlled I_E and, thus, I_C. The voltage amplifier works the same way, but with a different outcome. Consider the previous transistor arrangement from another perspective: We redraw it in the form shown in Figure 9.15. The change in base voltage is

$$\Delta V_B = \Delta V_E \tag{9.19}$$

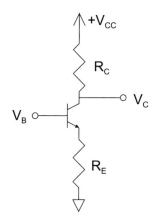

Figure 9.15. Transistor voltage amplifier.

Recall the analysis of Section 9.1.3: Constant terms are dropped when we consider changes. This way, even though V_B and V_E are related by a constant term (V_{BE}), the changes are not. That is, V_B and V_E increase or decrease by the same amount, and thus Equation 9.19 is true. Similarly, ΔV_E is also given by

$$\Delta V_E \simeq \Delta I_C R_E \qquad (9.20)$$

Since the voltage of the collector is

$$V_C = V_{CC} - I_C R_C \qquad (9.21)$$

then

$$\Delta V_C = -\Delta I_C R_C \qquad (9.22)$$

Combining Equations 9.19, 9.20, and 9.22, we get

$$\frac{\Delta V_C}{\Delta V_B} = -\frac{R_C}{R_E} \qquad (9.23)$$

That is, if $R_C > R_E$, the circuit behaves as an amplifier of AC voltages, with a gain of $-\frac{R_C}{R_E}$. (The negative sign means that the output is 180° out of phase with the input.) To use the transistor amplifier shown here we must bias the transistor, described next.

9.1.6   Biasing the Transistor

The versatility of operational amplifiers, to be discussed in the next chapter, has rendered biasing the transistor a thing of the past. The use of the transistor in the emitter-follower or the voltage amplifier as discussed in previous sections is limited by the restricted range of input voltage values that the transistor accepts: The input voltage must be fully above the forward drop of the transistor and below the supply voltage. Consequently, the input voltage needs to be an AC signal with a low-voltage swing above the voltage drop of the base-emitter junction. This limitation is overcome by biasing the transistor.

Figures 9.13 and 9.16 show the biasing arrangements for the two transistor circuits: emitter follower and voltage amplifier. We can summarize their differences and similarities the following way:

- Both circuits have an input capacitor C_1 that is used to take out any DC component of the input.
- The resistors R_1 and R_2 form a voltage divider that adds a DC voltage to the input. The combination of the capacitor C_1 and the input impedance of the voltage divider yield the cutoff frequency of the input high-pass filter, which must be *below* the frequency of the input. The values of R_1 and R_2 must be calculated so that the input and output voltages are not clipped.
- The output capacitor C_2 and the load resistor R_L form a high-pass filter that removes the artificial DC voltage that the transistor adds. The value of C_2 is calculated based on the signal frequency and R_L.

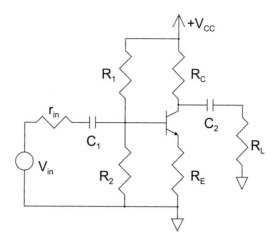

Figure 9.16.   Biasing of the voltage amplifier.

9.2 FIELD-EFFECT TRANSISTORS

To this point, we have studied bipolar-junction transistors (BJT), which are driven by a small base current. Field-effect transistors (FET) are transistors driven by a voltage. BJTs and FETs are built differently and thus have important differences. In terms of the direction of current flow, BJT transistors can be of two types: npn and pnp. FETs have a similar distinction and are referred to as n-channel and p-channel. In addition, FETs have three commonly used subtypes: enhancement-mode MOSFET, depletion-mode MOSFET, and JFET. Although subtle operational differences exist, the latter two are functionally the same.

9.2.1 Inside the FET

We start with the MOSFET, shown in Figure 9.18. The n-channel MOSFET has two electrodes connected to n-type regions: the drain (D) and the source (S). They are separated by a p-type region, called the body. The region of the body separating the two n-regions is connected to a thin insulator (silicon dioxide). At the other side of the insulator is the metal electrode, called the gate (G). The gate is electrically insulated from the drain, source and body.

In enhancement-mode MOSFETs, when no voltage is applied to the gate, no current can flow from the drain to the source. In essence, from drain to source, there are two p-n diode junctions back to back. A voltage on the gate above some minimum value V_T induces a piling of charge carriers at the oxide-body surface, also known as the inversion layer or channel, which allows a current to flow from drain to source. In the case of a body that is p-type, as shown in Figure 9.17, a positive gate voltage forms a channel where the carriers are electrons, which makes this an n-channel MOSFET. Conversely, if the body is n-type, a negative gate voltage forms carriers that are mobile holes. This is the p-channel MOSFET. In both cases, the gate controls the drain-source current with a voltage. This is different than the BJT transistor, in which a base current controls the collector-emitter current.

Because the gate is electrically insulated from the drain and the source, the gate has, in principle, infinite input impedance. In practice, it is about 10^{14} Ω, but this is quite high anyway. The value of V_T depends on the gate materials and dimensions. A value of V_T below zero results in a depletion-mode MOSFET. This is a MOSFET with a conducting

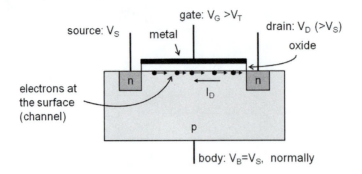

Figure 9.17. Schematic of layout and diagram of operation of the n-channel MOSFET.

channel already present at zero gate voltage, requiring a reverse gate voltage to actually turn off the device.

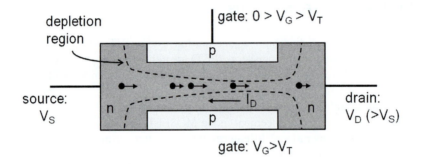

Figure 9.18. Schematic of layout and diagram of operation of the JFET.

The JFET is constructed differently, as shown in Figure 9.18. The drain and source are the ends of one continuous region. The gate is a doped semiconductor region of the opposite type than the source and drain. The two ends of the gate are electrically connected (the figure does not show this connection). The JFET also controls the drain-source current with a voltage at the gate, but because the gate-source/drain is a p–n junction, it cannot be forward biased. The JFET in the figure is an n-channel transistor, so the voltage of the gate (p-type) cannot be above the source (n-type) by more than 0.6 V. Otherwise, you will have an unwanted current from the gate to the source and no FET action. Thus, already at zero voltage, the JFET conducts. If the gate voltage is reduced *below* zero, a depletion region is created (the dashed line in Figure 9.18) that restricts the channel for passage of current from drain to source. At some value of the voltage, called the pinch-off voltage, the depletion regions touch, the channel gets completely pinched off, and no more current can flow. The gate then controls the flow of the current with a *negative* voltage.

9.2.2 Operation of the FET

Figure 9.19 shows the symbols for the different FETs. For n-channel FETs the current flows from the drain to the source, controlled by the gate. These are the homologous of the npn-BJTs collector, emitter, and base, respectively. The MOSFET has an extra connection, the body, which has to be reverse biased with respect to the source. Normally, it is internally connected to the source.

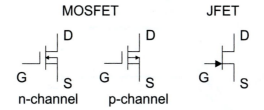

Figure 9.19. Symbols for the two types of FET transistors.

In an n-channel FET the drain is biased positive with respect to the source, just as the collector and the emitter are biased in a BJT. In the case of an FET, a drain current appears when the gate voltage exceeds a threshold V_T. For the case of the MOSFET, this threshold is positive and varies between 0.8 and 2.4 V, depending on the type of MOSFET. So at $V_{GS} = 0$, the MOSFET is off.

We can graph the FET curve in the same way as for the BJT. Figure 9.20 shows the operating curves for a MOSFET. They mimic the curves for the BJT: a fast rise followed by a flat region. When we blow up the early part of the characteristic curve (see Figure 9.21) of the FET, we see that the drain current increases linearly with the drain-to-source voltage. This is called the linear region. Where the drain current flattens is called the saturation region (here the name makes sense, but to avoid confusing it with the BJT, we will call it the FET saturation region).

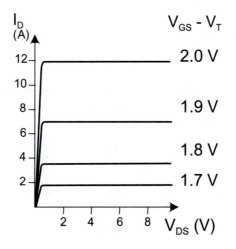

Figure 9.20. I–V curves for the FET transistor.

For a given value of $V_{GS} > V_T$, the current I_D increases nearly linearly with V_{DS}. We call V_{DSsat} the value of the drain-source voltage at the "knee," after which the FET enters the FET saturation region. When $V_{DS} < V_{DSsat}$, $I_D \propto V_{DS}$. Recall that the I–V curve for a resistor is a straight line that goes through zero with a slope that is the inverse of the

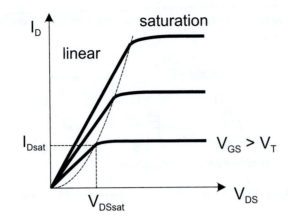

Figure 9.21. A close look at the turn-on section of the FET I–V curve.

resistance. We can then write the I–V relationship between the drain and source in Ohm's law fashion as

$$I_D = \frac{1}{R_{DS}} V_{DS} \qquad (9.24)$$

where R_{DS} is the FET resistance, given by

$$R_{DS} = \frac{1}{2k(V_{GS} - V_T)} \qquad (9.25)$$

Here, k is a constant, typically of about 0.3 mA/V^2, making R_{DS} about 3.3 kΩ-V. An interesting aspect of this equation is that R_{DS} depends on $V_{GS} - V_T$. This makes the FET a voltage-controlled variable resistor.

The value of the drain-source voltage at the knee is given by

$$V_{DSsat} = V_{GS} - V_T \qquad (9.26)$$

which is a peculiar result. It means that we are missing some information that leads to this relationship (a lot of physics!). When V_{DS} gets closer to the knee, the straight line starts to curve. A more accurate expression for the current to account for the nonlinearity is given by

$$I_D = 2k(V_{GS} - V_T)V_{DS} - kV_{DS}^2 \qquad (9.27)$$

We can make another set of relations that are satisfied by the knee of the curves. The coordinates of the knee are (V_{DSsat}, I_{Dsat}). We can use Equation 9.27 to obtain a relationship between V_{DSsat} and I_{Dsat}:

$$I_{Dsat} = kV_{DSsat}^2 \qquad (9.28)$$

That is, the coordinates of the knees follow a parabola, as shown in Figure 9.21.

When $V_{DS} > V_{DS(Sat)}$, the drain current stops rising and levels off. In BJTs, the value of I_C (for fixed V_{CE}) in the flat region of the transistor curve varied linearly with I_B ($I_C = \beta I_B$). For FETs the value of I_D (for fixed V_{DS}) in the flat FET saturation region depends on $V_{GS} - V_T$, not linearly, but quadratically:

$$I_D = k(V_{GS} - V_T)^2 \qquad (9.29)$$

(This equation is, of course, identical to Equation 9.28.) In summary, the important relationships for designing with FETs are Equations 9.24, 9.25, 9.26, and 9.29. In the following sections we discuss three applications of FETs.

9.2.3 The MOSFET Switch

One attractive feature of the MOSFET is that the gate can be driven above the drain voltage. This leads to a steep linear curve between I_D and V_{DS} in the region of operation. A steep slope means a small resistance, so the FET behaves like a small resistor. The left circuit in Figure 9.22 is the MOSFET version of the transistor switch that we saw in Section 9.1.2. Because of the voltage-driving capabilities of the FET, we no longer need to put a resistor in series with the switch as we did for the BJT (no current flows into the gate): We just drive it directly with a voltage.

The circuit on the right of Figure 9.22 is much more interesting: The MOSFET acts like a mechanical switch even when the drain voltage is lower than V_{GS}. When the drain voltage approaches V_{GS} ending the linear region, the resistance starts to increase and the MOSFET is no longer a practical switch. This use of the MOSFET is the basis of the 4066 IC that we mentioned in Section 1.11.1 to implement the "darkness sensor." In summary, for implementing the MOSFET switch, it is important to be in the linear region, or satisfying $V_{GS} - V_T > V_{DS}$.

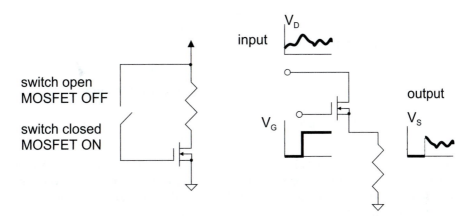

Figure 9.22. Circuits for two ways to use the MOSFET as a switch.

MOSFETs are also used for digital circuits. The switching characteristics are not very impressive because of the large capacitance involved. MOSFETs may have a short switch-on time, but they have a long switch-off time. To remedy this, the CMOS NOT gate consists of two MOSFET switches of different polarity (one n-channel and one p-channel) connected in the push-pull configuration, as shown in Figure 9.23. MOSFETs have important applications in power switching. Special high-voltage MOSFETs can be switched on with a relatively low voltage.

9.2.4 Current Sources

The best current sources are the ones driven by operational amplifiers, as we discuss in Chapter 10 (Operational Amplifiers). However, JFETs can make for simple ones.

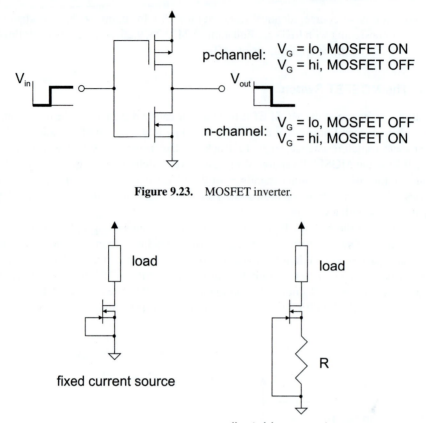

Figure 9.23.   MOSFET inverter.

fixed current source

adjustable current source

Figure 9.24.   FET current sources.

The one on the left of Figure 9.24 is worthy of notice because of its simplicity. Because the threshold voltage is negative, tying the gate to the source biases it for a set current rating. You are stuck with the value for the particular JFET, but these are normally sorted out and sold by their current rating. The current in the circuit on the right side of Figure 9.24 is adjustable. It uses the drop at R to set the gate bias and, consequently, the current. It is a form of self-biasing.

   Because of their high input impedance, FETs also find important applications in source followers and amplifiers. However, their use in discrete circuits is not common. They are used more efficiently in connection with operational amplifiers.

9.2.5 Variable Resistors

FETs find an interesting application as variable resistors. In the early region of operation, the drain current varies linearly with the drain-source voltage. Equation 9.25 gives the slope of the early region. Thus, by changing V_{GS}, we have a voltage-controlled resistor. Figure 9.25 illustrates this. The more accurate value for the resistance is given by

$$\frac{1}{R_{DS}} = 2k \left[(V_{GS} - V_T) - \frac{V_{DS}}{2} \right] \tag{9.30}$$

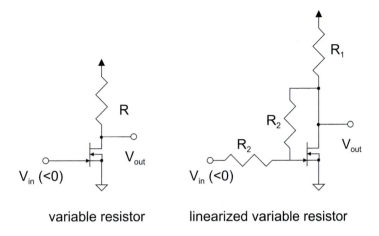

variable resistor   linearized variable resistor

Figure 9.25.   Circuits in which the FET is used as a variable resistor.

Power MOSFETs normally have a very small resistance, so they are used for switching. For example, IRF840 has an ON resistance of about 0.1 Ω for typical parameters. JFETs have much larger ON resistance. For example, 2N3819 JFET has an ON resistance of 150 Ω for typical operating parameters.

We can improve upon the circuit on the left side of Figure 9.25 by linearizing it, or eliminating the term $\frac{V_{DS}}{2}$ in Equation 9.30. Consider the circuit on the right side of Figure 9.25. Because the source is grounded, $V_{DS} = V_D$. The current flowing to the input is

$$I_{in} = (V_D - V_{in})/(2R_2).$$

The gate voltage then is

$$V_G = R_2 I_{in} + V_{in}.$$

■ **EXERCISE 9.6**

Use Equation 9.27 to show that the drain current becomes

$$I_D = 2k(V_{in} - V_T)V_D$$

The two resistors R_2 linearize the relationship between I_D and V_D, making the FET behave *exactly* like a resistor.

9.3 PROBLEMS

1. Calculate the value of R_B for the circuit of Figure 9.26 so that it behaves as a transistor switch.

2. Figure 9.5 gives the I–V curve of the transistor in the circuit of Figure 9.27. What is the current flowing through the load?

3. The input waveform in the circuit of Figure 9.28 is $V_{in} = V_0 \cos \omega t$, with $V_0 = 3$ V and a frequency $f = 10$ kHz.

    (a) Make a sketch of V_E.

    (b) If $R_E = 100$ Ω, sketch the current flowing through it. Label your axes.

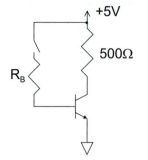

Figure 9.26. Circuit for Problem 1.

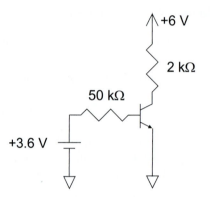

Figure 9.27. Circuit for Problem 2.

(c) Suppose that we now add a DC component to the input so that

$$V_{in} = V_0 \cos \omega t + V_{DC}$$

Sketch the output waveform of V_E for the following values of V_{DC}: 1 V, 3 V, and 5 V.

(d) With $V_{DC} = 5$ V, attach a filter to the output, with a resistor $R_L = 1$ kΩ and a capacitor C_2. If we want to suppress only the DC part of V_E, with at least 90 percent of the 10-kHz waveform coming out, what value of C_2 should we use? Draw a circuit diagram.

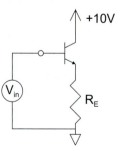

Figure 9.28. Circuit for Problem 3.

4. For the circuit of Figure 9.29:

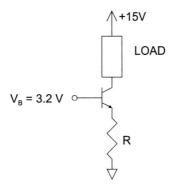

Figure 9.29.   Circuit for Problem 4.

   (a) Determine the value of the resistor R so that a current of 0.5 A flows through the load.

   (b) What is the maximum value of the load resistor?

   (c) If the load resistor is 10 Ω, what value of V_B will put the transistor into saturation?

5. What is the relationship between the input voltage V_B and I_C in the current source circuit shown in Figure 9.30? What is the maximum current that can flow?

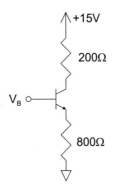

Figure 9.30.   Circuit for Problem 5.

6. Design a 0–10 mA current source.

   (a) If the load is 1 kΩ, what is the maximum drop across the load?

   (b) Pick a value of V_{cc} that is greater than the voltage you calculated in the previous part.

   (c) Explain why we should do the latter.

   (d) What value of R_E would you pick?

   (e) Give a relationship between V_B and the current flowing through the load.

7. Design a current source that can put out a current of up to 50 mA to a maximum load of 500 Ω. Assume that $V_{CC} = 30$ V.

8. We apply 2 V to the base of the transistor, as shown in Figure 9.31. If $R_E = 1,000\ \Omega$ and $R_C = 2,000\ \Omega$, what is the voltage of the collector of the transistor relative to ground?

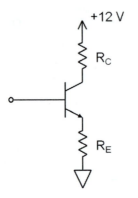

Figure 9.31.   Circuit for Problem 8.

9. The transistors shown in Figure 9.32 have $\beta = 100$.

   (a) Find the current flowing through the 4-Ω resistor.
   (b) Find the current flowing out of the 6-V source.
   (c) The output impedance of the 6-V source is 10 Ω. Is the circuit loading the source? Explain.

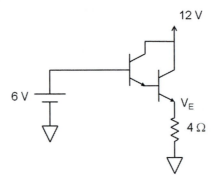

Figure 9.32.   Circuit for Problem 9.

10. An n-channel MOSFET has the following operating parameters: $V_T = 4\ \text{V}$ and $k = 2.8\ \text{A/V}^2$.

    (a) Design a transistor switch that applies 500 V to a 1 kΩ load.
    (b) What is the resistance of the MOSFET when $V_{GS} = 6\ \text{V}$ and $V_{DS} = 1\ \text{V}$?

11. The circuit of Figure 9.33 is a simple MOSFET NOT gate. The MOSFET parameters are $k = 0.2\ \text{mA/V}^2$ and $V_T = 0.8\ \text{V}$.

    (a) If the input is $V_{in} = 5\ \text{V}$, find I_{Dsat}.
    (b) Is the MOSFET in the linear region? Justify your answer.
    (c) Find V_{out}.

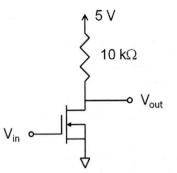

Figure 9.33.   Circuit for Problem 11.

12. Consider a JFET that has the curves shown in Figure 9.34.

   (a) Estimate the value of the threshold voltage V_T. Justify your answer.
   (b) For what value of the gate-source voltage is the current 2.2 mA?
   (c) For what values of the drain-source voltage is the current 2.2 mA?
   (d) What current flows through the circuit of Figure 9.35 when it is in the flat (FET saturation) region?

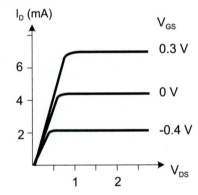

Figure 9.34.   FET current source for Problem 12.

   (e) If the FET is in the circuit of Figure 9.35, for what values of the drain-source voltage is the FET in the flat region? Justify.

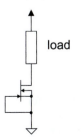

Figure 9.35.   Circuit for Problem 12(d).

   (f) If the supply voltage is 3 V, and the load is 375 Ω, use the load-line method to determine whether the FET is in the flat region.

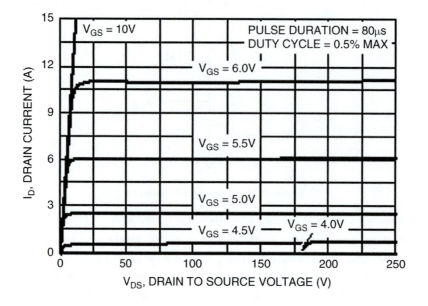

Figure 9.36.   I–V curve of the IRF840 MOSFET (with permission from Fairchild Semiconductor, Inc.).

    (g) Suppose that the supply voltage is set to a value for which the FET is in the linear region. In this region, it acts like a resistor. What is its resistance?

13. For the MOSFET of Figure 9.36,

    (a) Find the threshold voltage.

    (b) If we have a 100-V power supply, for what range of loads can the MOSFET be used as a switch with $V_{GS} = 6$ V?

    (c) Estimate the ON resistance for $V_{GS} = 6$ V.

9.4 LAB PROJECTS

This section presents ideas for lab projects.

9.4.1 BJT Transistors

Transistor's Curve

Required equipment: Curve tracer.
Required components: 2N3904 npn transistor.

1. Use the curve tracer to obtain the characteristic curve of the 2N3904 npn transistor (see Figure 9.37). From the transistor curve(s), obtain an approximate value for β.

2. On your transistor curve, draw a load line corresponding to $V_{CC} = +12$ V and $R_C = 1200\ \Omega$. If $I_B = 30\ \mu A$, what are the values of V_{CE} and I_C? Set up the circuit shown in Figure 9.38 and check your predictions.

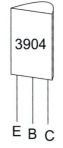

Figure 9.37. Pin connections of the npn transistor 3904.

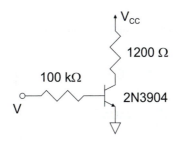

Figure 9.38. Circuit for lab project on BJT transistors.

Application: Transistor Switch

Required equipment: Power supply.
Required components: 2N3904 npn transistor, 12 V relay, LED, resistors.
Suppose that we want to turn on a high-power circuit. This is best done using a relay as the ON switch. A relay separates the high-power circuit from the driving circuit, leaving the more modest task of turning on the relay to our circuit. Suppose, however, that we have a limited source to turn on the relay: a 2-V power supply. A solution is to use another switch, a transistor switch, to drive the relay. As shown in the circuit of Figure 9.39, we have a 12-V relay. As a high-power load, we just have an LED, but we could easily have any other device that we want to power.

1. We want to use the transistor as a switch. The load is the relay. How much current flows through the relay when you drive it with 12 V?

2. Using the curve tracer, take a set of transistor curves so that you can draw a load line in the graph. From the load-line analysis, estimate the value of base current needed.

3. Find the value of R_B.

The Emitter Follower

Required equipment: Power supply, oscilloscope, function generator, audio speaker.
Required components: 3055 npn transistor, 2955 pnp transistor, 100 μF capacitors. The transistors may have to be connected to heat sink elements to dissipate heat.

1. Set up the function generator to 0.2 V peak, 1 kHz and a DC offset of 0 V. Connect the function generator to the speaker through the emitter-follower circuit shown in Figure 9.40. View the input and the output signals in channels 1 and 2 of the scope.

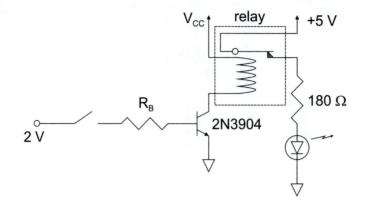

Figure 9.39. Circuit for lab on the transistor switch.

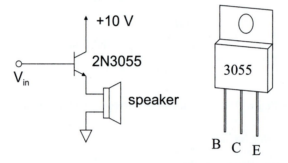

Figure 9.40. Wiring of circuit for lab on power transistor 3055.

Write down and explain what you observe as you slowly increase the DC offset of the generator to 1 V. Make a sketch. Can you explain all the features of the waveform you see?

2. Reduce the DC offset of the supply to zero and assemble the circuit shown in Figure 9.41. View the input and output signals in channels 1 and 2 of the scope. Write down and explain what you observe when you increase the amplitude of the voltage to 2 V. Do this momentarily, or the transistors will heat up quickly. State the benefits and drawbacks of this circuit. Note that the pin connections of the 2955 and 3055 transistors are the same (see Figure 9.40).

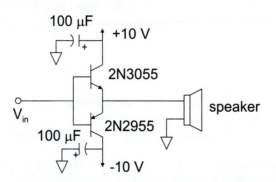

Figure 9.41. Circuit on the transistor push-pull.

Current Source

Required equipment: Power supply, multimeter.
Required components: 2N3904 npn transistor, resistors, potentiometer.

1. Design a current source that supplies a 0–10 mA current to a 500 Ω load, as shown in Figure 9.42. Find the value of V_E such that the transistor is below saturation when supplying 10 mA (use your judgment). Knowing V_E, you can calculate V_B and R_E.

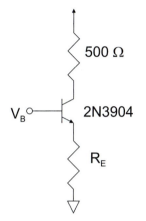

Figure 9.42.   Circuit of lab project on the current source.

2. Instead of using a separate supply to generate V_B, generate it by voltage-dividing the supply voltage with a "pot" (short for potentiometer), as shown in Figure 9.43. The pot consists of a resistor with a variable middle contact. Turning the pot dial changes the position of the middle contact. When applying the supply voltage to the extremes of the pot, the middle contact is at a potential determined by its position, acting as a voltage divider. You should choose one of three sizes of pots (specified by the value of its total resistance): 1 kΩ, 10 kΩ, or 50 kΩ. The criterion is that the transistor should not load the divider.

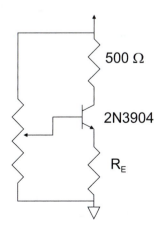

Figure 9.43.   Circuit for lab project on the current source.

3. Assemble the circuit with the pot set to zero. As you increase its value, take a few measurements of I_C vs. V_B. Make a plot and find the transconductance of the amplifier.

9.4.2 FET

FETs I–V Curve

Required equipment: Curve tracer.
Required components: JFET MPF102.

1. Using the curve tracer, obtain the characteristic curve of the MPF102 FET (see Figure 9.44).

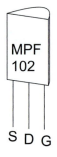

MPF
102

S D G

Figure 9.44.   Pin connections for the FET MPF102.

2. From the transistor curves and Equation 9.26, determine V_T.

3. Obtain an approximate value for k.

Application: AM Transmission with FET As a Variable Resistor

Required equipment: Two function generators, oscilloscope, AM radio.
Required components: JFET MPF102 (or equivalent), resistor, long wire serving as an antenna.

Consider the circuit shown in Figure 9.45. Connect the function generator with a 1 MHz 1-V input signal.

Connect the channel 1 of the scope to the 1-MHz supply and channel 2 to the drain. When V is less than V_T, the FET is an open circuit. When V_G is increased to be greater than V_T, the FET starts to conduct. The larger the value of V_G, the steeper the linear region

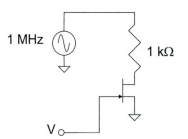

1 MHz

1 kΩ

V

Figure 9.45.   Circuit for testing the FET as a variable resistor.

and thus the smaller the effective resistance of the FET. Eventually, the FET enters the FET saturation (flat) region, and it stops acting as a variable resistor.

1. Verify that the 1-kΩ resistor and the FET work like a voltage divider controlled by the gate voltage. Correlate the amplitude of the drain voltage with V_G.

2. Now, instead of using a constant voltage, connect the gate to the second function generator with a 1-kHz waveform that oscillates between $V_T + 1$ V and $V_T + 2$V. Observe V_{DS} and V_{GS} in the scope, triggering on V_{GS}. Make a sketch of the waveform and explain it.

3. Broadcast to an AM radio located in the room, tuned to 1 MHz. Vary the frequency of V_{GS}. Attach a long wire to the drain of the FET to serve as an antenna.

MOSFET Switch

Required equipment: Function generator, power supply.
Required components: MOSFET IRF840, resistor.

We want to switch an input voltage of 1 V peak to peak, a frequency of 1 kHz, and a DC component of 1 V, through a MOSFET and into a 1 kΩ load (see Figure 9.46). Use the IRF840 MOSFET for this. Figure 9.36 shows the data curves.

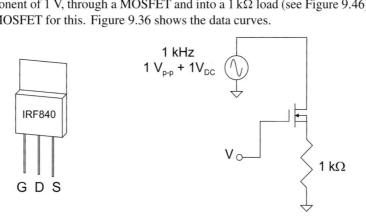

Figure 9.46.   Circuit for testing the MOSFET switch with the MOSFET IRF840.

1. Observe both input and output as a function of the gate voltage. Record your observations.

2. From your measurements, estimate the on resistance of the MOSFET for a gate voltage of 7 V.

9.4.3 Practicum Test

1. Use the function generator to generate a triangular signal with an amplitude of 1 V and a frequency of 2 Hz.

2. Connect the emitter of a transistor (3055) to a small incandescent light bulb, as shown in Figure 9.47.

3. Apply the waveform from the first part to the base of the transistor. Make a graph of the current flowing through the light bulb. (Label your axes!) Your graph must be consistent with your measurements.

4. Use the DC component of the function generator to make the bulb stay on permanently.

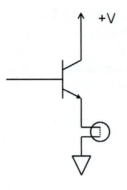

Figure 9.47.   Circuit for the practicum.

CHAPTER 10

OPERATIONAL AMPLIFIERS

Contents

This chapter is devoted to perhaps the essence of electronics: amplification. We do something, we sense something, and then, unavoidably, we must amplify it. As we interface with the real world, we get all kinds of inputs. Most of them are small, maybe even imperceptible to simple measuring devices. If we want to do something with those inputs, we must make them bigger by amplifying them. In the previous chapter, we saw that a transistor can be used to amplify. Indeed, the transistor *is* used to amplify. A discrete transistor, however, is very raw; we must do a lot of designing to get what we want. The operational amplifier, or op-amp for short, is a clever circuit with a long history. It started out with vacuum tubes, but the advent of the integrated-circuit operational amplifier more than 50 years ago made a huge difference in the way we approach electronics. Op-amps are differential amplifiers, which sounds like a difficult concept, but it is not. They amplify a difference. Op-amps are not ready-to-use boxes. Rather, they are the main building block of analog circuits. Thus, we need to add components to implement the application of our choice. If this circuitry includes negative feedback, we obtain marvelous circuits.

Beyond their functionality, today's op-amps really add to the pleasure of doing electronics. They operate under simple rules, they are easy to wire, and the results are immediate. You cannot ignore the previous chapters, but compared to them, this topic is a joy ride. In difficulty, the course just went over the mountain range and is going down to the plains. Although the previous chapter was at the peak of difficulty, you cannot ignore transistors. We use them in combination with op-amps to deliver power because op-amps do the finesse but not the brute force. We start with the basics, and you will see that this chapter is the highlight of the course. We end the chapter by coming back down to earth in terms of their limitations and the details we need to consider to get the results that we want.

10.1 NEGATIVE FEEDBACK

Op-amps are most useful when used in conjunction with negative feedback. This is a simple but profound concept: You subtract from the input a sample of the output. This seems ludicrous. Indeed, Harold Black's idea in 1934 was treated with as much ridicule as Robert Goddard's after he invented the multistage rocket for space exploration. But here is the importance of negative feedback: It renders a nonlinear device linear. Humans and animals hugely depend on negative feedback and corrective actions to walk, run, go straight, or even stand upright. Our senses provide the feedback, and our brain circuitry lets those action potentials flow through nerves to unleash muscles of all kinds. We do not really need to think to be able to lift our hand to where we want. Proof of this is that most animals, smart or not, are well wired to move and fly by instinct. Suppose that you want to touch your nose. Feedback lets your muscular arm rush up to your face. Visual feedback allows you to slow it down appropriately, and the sense of touch makes your finger gently land on your nose, not too hard and not too soft, so it feels just right. This is all feedback. Put a broom upright and let it go: It swings right back down. The broom does not have a mechanism that compensates, so that it changes, leans, and flexes to remain upright. How do we do it? We look and we feel, and that information is sent to our muscles and extremities as a corrective action. Op-amps provide the simplest building blocks of negative feedback, to provide stability, amplification, and much more. So let us see how op-amps use negative feedback to do their magic.

A differential amplifier takes the difference of two inputs, V_+ and V_-, and amplifies it by an amount G that we call the *open-loop* gain:

$$V_{out} = G(V_+ - V_-) \tag{10.1}$$

Op-amps are differential amplifiers with a high open-loop gain. We represent them in the block form shown in Figure 10.1.

Not shown are other inputs, such as power connections, but we get to them later. The gain G is not a reliable parameter; it depends on the frequency, and we may not have any control over it. With negative feedback, we can design a circuit with a gain that is constant and that we can control. We do this by feeding the *inverting* input (V_-) with a sample of the output.

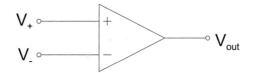

Figure 10.1. Signal connections in an op-amp.

Let us see how this stable gain is accomplished. Suppose that we connect an input voltage V_{in} to the *noninverting* input of the op-amp (V_+), and a fraction $\frac{1}{A}$ of the output to the inverting input (V_-), as shown in Figure 10.2. Then Equation 10.1 becomes

$$V_{out} = G\left(V_{in} - \frac{1}{A}V_{out}\right) \tag{10.2}$$

Solving for the output voltage, we get

$$V_{out} = \frac{G}{1 + \frac{G}{A}}V_{in} \tag{10.3}$$

If $G \gg A$, then

$$V_{out} \simeq AV_{in} \tag{10.4}$$

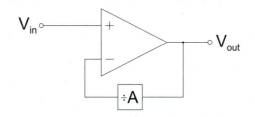

Figure 10.2. Negative feedback with an op-amp.

Equation 10.4 states that the output voltage is equal to the input voltage multiplied by a gain A. It is attractive for its simplicity and because we can control the value of A. The *closed-loop gain A* will be smaller than G but independent of it. This simple principle makes operational amplifiers one of the most useful analog electronics ICs.

This chapter is devoted almost exclusively to op-amps that need two power supplies (dual-supply op-amps). However, new trends in op-amp development focus on op-amps with a single low-voltage supply. These can easily interface with portable electronic devices that are battery operated. Near the end of the chapter, we devote a section to single-supply circuits.

Here we list a few important practical features of op-amps:

- Figure 10.3 shows the typical wiring of a dual-supply op-amp, with power supply voltages V_{CC} and V_{EE}. The signal inputs, V_1 and V_2, feed into the boxes labeled "circuit," which contain circuit components related to the design. The figure also shows the pin connections of an ordinary op-amp. One of the more puzzling issues for a beginner is the ground connection of the op-amp. Where is it? The ground level is *implicit* because the input signals are referenced to a common line that we often designate as ground, as shown explicitly in the figure. Make sure that the grounds of the inputs and the power supplies are all connected together. The op-amp will "know" where ground is. The output will be referenced to the same common ground.

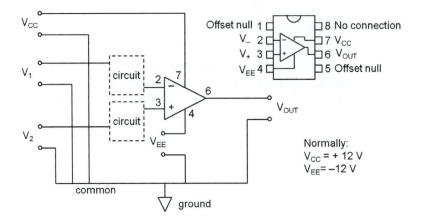

Figure 10.3.   Diagram explicitly showing the power, signal and common-ground connections.

Figure 10.3 explicitly shows the common wire and the two supplies, but you can see that the circuit is cluttered with wires and two-terminal inputs and outputs. For the sake of simplicity, beyond this point, we omit those common wires and supply lines unless they are necessary. The notation is also typical of electronics literature, so be sure to learn to read and write in electronics jargon. In the end, a bit of abstraction goes a long way for the sake of simplicity.

A note is needed on the wiring of the supplies, a topic that also confuses beginners. The op-amps are powered by positive and negative supplies, generally labeled as $V_{CC} = V_{CC+} = +V_S$ and $V_{EE} = V_{CC-} = -V_S$ in manufacturers' data sheets. Typically, V_S ranges between 12 and 15 V. If we have two power supplies with ungrounded outputs, we can set them both to, say, 12 V and connect them as shown in Figure 10.4.

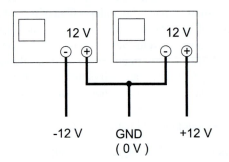

Figure 10.4.   Power connections for dual-supply op-amps.

- The inputs + and − of the op-amp are connected to the base of BJTs or to the gates of FETs. Thus, they draw very little current. For now, let us assume that they draw no current; they just sense the voltage applied to them. In precision circuits, we need to look at the specifications regarding this issue.

- The output amplifies a difference at the input, as given by Equation 10.1. Because open-loop gains can be as high as 10^6, a small difference at the input makes the output voltage increase to its maximum value of $\pm V_S$, also known as the rail (the name may tell you how circuits were wired in the old days). Note also that this gain depends on the frequency of the input signal, decreasing as we go to higher frequencies, and becoming only 1 around 1 MHz.

- Most op-amps cannot source or sink large currents, so plan on getting *no more than tens of milliamperes* from them. In Section 10.2.2 we discuss options for enhancing their power capabilities.

This is all we need to know for now. In the next two sections, we apply the simple and ideal operating principles of op-amps for two configurations: closed-loop and open-loop. We will see many interesting circuits. It is a lot of good stuff, and a lot of fun to figure out. In the third section of the chapter, we give an overview of the limitations and other aspects of real op-amps. Precision circuits have to take those specifications seriously. In this course, we may be able to get away without them, but in the future, you will have to worry about op-amps' dirty little secrets.

10.2 CLOSED-LOOP CIRCUITS

Horowitz and Hill[1] have two simple rules for closed-loop op-amp amplifiers in their classic text (for many, *the* electronics text—if you plan to do serious electronics, you must get it!). They call them the *Golden Rules*:

   I. The op-amp sets the output voltage V_{out} so that the difference between two input voltages V_+ and V_- is zero.

   II. The inputs draw no current.

These two simple rules are all we need to know to understand and design simple op-amp circuits. In some ways, the op-amp brings back some memories of digital electronics, where gates were nice black boxes that did their job without making us worry about the electronic horrors inside. Next, we cover a nice variety of useful and interesting circuits.

10.2.1 Noninverting Amplifier

A first example of negative feedback is the noninverting amplifier. As shown in Figure 10.5, V_{in} is connected to the noninverting input. The inverting input samples the output via a voltage divider so that

$$V_- = \frac{R_1}{R_1 + R_2} V_{out} \tag{10.5}$$

Let us see how this works. Suppose that, initially, V_{out} is zero. Since $V_{in} > 0$, then there is an imbalance between V_+ and $V-$ (that is, $V_+ > V_-$), and V_{out} grows until V_- reaches V_+.

[1]P. Horowitz and W. Hill, *The Art of Electronics*, 2nd ed. (Cambridge: Cambridge University Press, 1989). Note also that this text has a huge list of op-amps of different kinds, so we avoid listing them given that this reference has done it so well.

If V_- goes over, then V_{out} starts to decrease. In the end the circuit stabilizes when $V_+ \simeq V_-$. For design purposes, assume that

$$V_+ = V_- \tag{10.6}$$

However, in reality, they differ by a very small amount: $\frac{1}{G}V_{out}$. Rearranging Equation 10.5 and combining it with Equation 10.6, we get

$$V_{out} = \left(1 + \frac{R_2}{R_1}\right) V_{in} \tag{10.7}$$

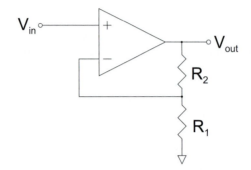

Figure 10.5. Noninverting amplifier.

The closed-loop gain is then $A = 1 + \frac{R_2}{R_1}$. This amplifier has a few interesting aspects. Above all, its simplicity makes this an appealing device. Unlike with transistors, we do not need to worry about biasing. As shown in the previous design, we need to specify only two resistors. We still need to worry about limits and other details, but those are issues we need to worry about with any circuit. Another good feature of this amplifier is that it amplifies any voltage, positive or negative, constant or variable, as long as the output does not exceed the power supply voltage. As if that is not enough good news, it also has a *very* high input impedance. Thus, we do not have to worry about loading.

You may be wondering, where does the current come from? Remember that the op-amp *is* a circuit, which draws power from external supplies that *must* be connected. Moreover, if V_{out} is negative, the op-amp *draws* current into it. That is okay, too, because one of the supplies of the op-amp is negative. Remember that a lot is hiding behind that simple triangular symbol.

10.2.2 Follower

The thought that the noninverting amplifier has good impedance-matching properties raises the following question: Can we use it exclusively for that purpose? Sure! You only need to make $R_2 = 0$ and $R_1 = \infty$, which yields $A = 1$, or $V_{out} = V_{in}$. The circuit then looks like the one shown in Figure 10.6.

This is also called a buffer: Because it has very large input impedance ($10^{12}\Omega$ for FET-based op-amps and $10^8\Omega$ for BJT-based op-amps), it can prevent loading. Op-amps have low output impedance (a few tens of an ohm) but can source only a few milliamperes. This limits the effectiveness of the buffer when driving low-impedance loads.

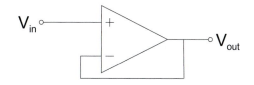

Figure 10.6. Op-amp follower.

Figure 10.7 shows a solution to the problem of the op-amp not being able to drive low-impedance loads This circuit uses a transistor to pull the raw power from the supply. It is also known as the power booster.

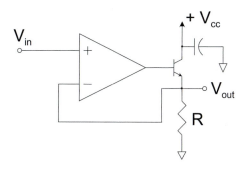

Figure 10.7. Op-amp power booster.

Remember that the op-amp increases V_{out} until $V_+ = V_-$. Thus, it drives the transistor in such a way that the current that flows through the load R creates a potential equal to the input voltage. There is no need to worry about V_{BE} because the feedback senses the emitter potential, not the base potential. The transistor provides the current that passes through the low-impedance load R. Because the op-amp drives the base of the transistor, it needs to supply only a small current to do the job.

You can also see that the previous circuit has a limitation: The transistor current flows in only one direction, so it works only for positive voltages. We solve this by using the push-pull in the circuit of Figure 10.8.

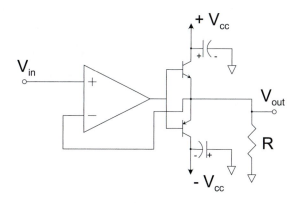

Figure 10.8. Op-amp power booster that swings in both polarities.

■ **EXERCISE 10.1**

The transistors in the previous circuit act as a current-supply extension of the op-amp. Without adding another op-amp, modify the circuit so that it also amplifies by a factor of five.

10.2.3 Inverting Amplifier

Did you notice a limitation of the previous type of circuit? Its amplification cannot be less than 1. In some situations, we might want to fine-tune a signal by making it smaller. In that case, we need an inverting amplifier, shown in Figure 10.9. Because $V_+ = 0$, V_{out}

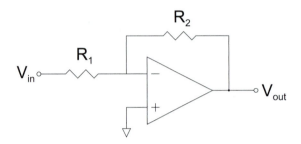

Figure 10.9. Inverting amplifier.

will decrease (if $V_{in} > 0$), and become negative, to make $V_- = 0$. If you think you can apply a simple voltage-divider formula to get V_{out}, think again. So far, we have dealt with the voltage divider when one end of the circuit is grounded. To keep it simple, we forget about the voltage divider for now.

A current

$$I = \frac{V_{in}}{R_1} = -\frac{V_{out}}{R_2} \tag{10.8}$$

will flow through R_1 and R_2 (remember that no current flows into the V_- input). The output voltage is given by

$$V_{out} = -\frac{R_2}{R_1} V_{in} \tag{10.9}$$

The negative sign means that the output is of the opposite polarity as the input. *The negative feedback will work as long as the feedback is connected to the inverting input (−).* If we connect it to the noninverting input (+), we will have a voltage runoff: The output voltage will increase because of a difference in the input, but the feedback will increase the difference between V_+ and V_- instead of decreasing it. Positive feedback can be used for a productive purpose but we discuss this later in Section 10.3.4, when we cover oscillators.

Now let us take a different view of the circuit by applying the voltage divider concept. Suppose that $V_{in} > 0$. If we apply the voltage divider concept, the low-voltage point is V_{out}. Therefore, all voltages must refer to it:

$$(V_- - V_{out}) = \frac{R_2}{R_1 + R_2}(V_{in} - V_{out}) \tag{10.10}$$

You can verify this for yourself. If we assume that $V_{in} < 0$ and refer to the voltages with respect to V_{in}, we get the same answer. You can verify yourself that when $V_- = 0$, Equation 10.10 leads to Equation 10.9.

■ **EXERCISE 10.2**

Derive an expression for V_{out} when the noninverting input of the previous circuit is connected to a potential V_0 (do not underestimate this calculation!).

If $V_{in} > 0$, then V_{out} is at a lower potential. A current then flows from the input, through the resistors, to the output. What then? Where does it go? As mentioned earlier, it goes into the op-amp.

10.2.4 Summing Amplifier

What more good news can we get? We are just getting started. Consider the circuit shown in Figure 10.10. We can assume that V_{out} will end up with a value that ensures that $V_+ = V_-$.

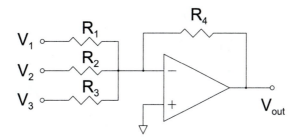

Figure 10.10.   Summing amplifier.

These currents flow through each of the input resistors:

$$I_1 = \frac{V_1}{R_1} \tag{10.11}$$

$$I_2 = \frac{V_2}{R_2} \tag{10.12}$$

$$I_3 = \frac{V_3}{R_3} \tag{10.13}$$

They converge to the feedback resistor R_4, so

$$I_4 = I_1 + I_2 + I_3 \tag{10.14}$$

with

$$I_4 = -\frac{V_{out}}{R_4} \tag{10.15}$$

Combining the previous equations, we get

$$V_{out} = -\left[\left(\frac{R_4}{R_1}\right)V_1 + \left(\frac{R_4}{R_2}\right)V_2 + \left(\frac{R_4}{R_3}\right)V_3\right] \tag{10.16}$$

The circuit amplifies the weighted sum of three input voltages, with each weight determined by the corresponding input resistor.

Note that the previous equations assume that $V_+ = 0$. Otherwise, we would have, for example,

$$I_1 = \frac{V_1 - V_+}{R_1}$$

and

$$I_4 = \frac{V_+ - V_{out}}{R_4}$$

The previous analyses assume that the op-amp inputs do not sink any current and that its output can sink or source current depending on the value of V_{out}. For most applications, that is still a good enough approximation. Later in Section 10.4, we discuss finer corrections that may be needed in more sophisticated circuits.

10.2.5 Differential Amplifier

The name is already giving it away: The circuit of Figure 10.11 amplifies the difference in two input voltages. Note that now $V_+ \neq 0$ is part of the design.

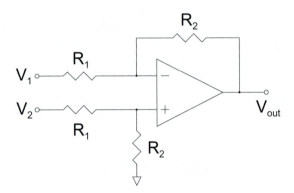

Figure 10.11. Differential amplifier.

■ **EXERCISE 10.3**

Derive the equation for the circuit of Figure 10.11 in the following steps:

1. *Find an expression for V_- in terms of V_1, V_{out}, R_1, and R_2.*
2. *Find an expression for V_+ in terms of V_2, R_1, and R_2.*
3. *Using the previous results, show that*

$$V_{out} = \frac{R_2}{R_1}(V_2 - V_1) \tag{10.17}$$

The differential amplifier of Figure 10.11 has the drawback that the input impedance may be too low for some applications. The differential input resistance is also low ($2R_1$). We can correct for this by using the instrumentation amplifier (see Problem 9 in connection with the figure on page 147).

10.2.6 Current Source

We can create current sources using the same trick as transistors: We apply a voltage over a known resistor. If we use noninverting amplifiers, then V_{in} controls the value of the current via V_- applied to a fixed resistor R so that

$$I = \frac{V_{in}}{R} \tag{10.18}$$

Circuit (a) of Figure 10.12 is the low-current version of the current source. The high-current version in (b) uses a power booster to provide higher currents because op-amps can go up to only tens of milliamperes. It could use an FET instead of the bipolar-junction transistor.

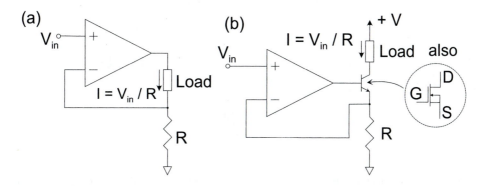

Figure 10.12.   Diagram of the use of an op-amp to make a current source (a), and its high-power version (b).

Note, however, that the load is not grounded. The ungrounded load may be inconvenient for some applications. Figure 10.13 shows a solution. It uses a pnp transistor. Remember that this transistor does the opposite of the npn transistor, so current goes from emitter to collector and the base sources a current, in our case, to the op-amp. A p-channel FET can replace the pnp transistor as well.

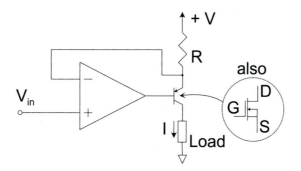

Figure 10.13.   Diagram of the circuit for making a current source with a ground connection (see also Problem 6).

The circuit of Figure 10.13 does not present the best situation because the input voltage is referenced not to ground, but to V_{CC}, so the current increases when V_{in} decreases. The source voltage then has to be well regulated and constant. Figure 10.14 shows a solution to this: A transistor converts a near-ground input voltage to the needed near-V_{CC} input to the current source. In essence, one current source drives another one.

■ **EXERCISE 10.4**

Show that the current flowing through the load in the circuit of Figure 10.14 is given by

$$I = \frac{R_2}{R_1 R_3} V_{in}$$

(10.19)

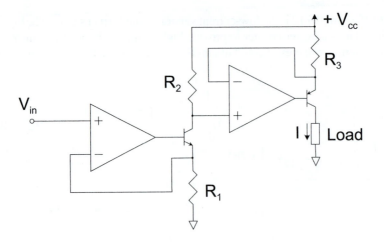

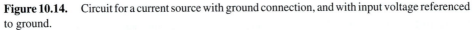

Figure 10.14. Circuit for a current source with ground connection, and with input voltage referenced to ground.

An important application of a current source is the generation of voltage references. In many circuits, we need a stable voltage to use as a reference. We know that the Zener diode has a sharp reverse-bias drop at the Zener voltage, but the drop does not have an infinite slope (it is typically 2 Ω). Thus, variations in the current flowing through the Zener cause the Zener voltage to change slightly. If a current source drives the Zener diode, and if the diode voltage is sampled with a high-impedance follower (an op-amp with FET inputs, as shown in the Figure 10.15), then the output voltage is very stable.

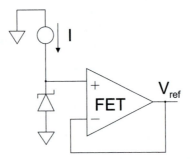

Figure 10.15. A voltage reference uses the best of what we have learned: a stable current source driving a Zener diode, with the Zener potential being buffered by an op-amp follower.

10.2.7 Current-to-Voltage Converter

The circuit of Figure 10.16 gives an output voltage V_{out} proportional to the input current I_{in} given by

$$V_{out} = -I_{in}R \qquad (10.20)$$

where I_{in} is defined as positive when flowing toward the op-amp.

The current source must count on the input being pulled to ground by the op-amp. Typical applications are photodetector circuits: Photodiodes and phototransistors are basically current sources, so the op-amp current-to-voltage converter is the ideal application, as shown in

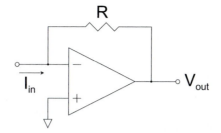

Figure 10.16.   A current-to-voltage converter.

Figure 10.17. As you remember from Chapter 8 (Diodes), a photodiode generates a reverse current when it detects light. The diode junction must be reverse biased or at zero potential to be most useful in the fourth quadrant of the diode curve. In the case of the phototransistor, the photo-current is a base current that is amplified by the transistor.

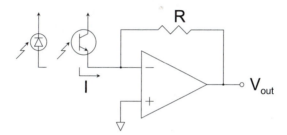

Figure 10.17.   Circuit to convert a photo-current from a photodiode or photo-transistor into a voltage.

10.2.8   Integrator

Consider the circuit of Figure 10.18. The current flowing into the capacitor is

$$I = \frac{V_{in}}{R} = -C\frac{dV_{out}}{dt} \qquad (10.21)$$

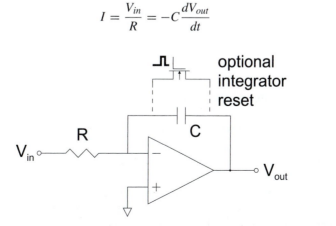

Figure 10.18.   Circuit for using an op-amp as an integrator.

Solving for V_{out}, we get

$$V_{out} = -\frac{1}{RC} \int V_{in} dt \qquad (10.22)$$

Thus, the output is a running time integral of the input. Check the interesting use of this circuit in connection with the lab project on servo-mechanisms (see Section 10.6.4).

If we want to have a way to reset the integrator, we can add a MOSFET with the source and drain connected to the ends of the capacitor, as shown in Figure 10.18. A pulse applied to the gate of the MOSFET shorts the capacitor and resets the integrator.

An Application of an Integrator: An Electrometer   We can use the integrator to make a handy electrometer. Suppose that we connect the input of the integrator and common lead to a capacitor with a charge Q. The op-amp then sets V_{out} to discharge that capacitor and set the inverting input to ground potential. In the process the integrator's capacitor is charged, making $V_{out} = \frac{Q}{C}$. If you use $C = 0.1\ \mu F$ and work out the units, it comes out to 10 mV/nC. You can then read the charge with a standard digital multimeter. For this purpose, we need to use a specialty op-amp: one with FETs in the inputs and very high input impedance, such as the LMC6001. If we pick $R = 10\ M\Omega$, it gives the circuit a time constant of 1 s. You should also connect a similar resistor between the noninverting input and ground to keep that input discharged. Instead of a MOSFET, we can use an ordinary switch across the capacitor.

10.2.9   Differentiator

At this point, we are figuring out all that we can do with op-amps. In the differentiator (see Figure 10.19), the current charging the capacitor is $C\frac{dV_{in}}{dt}$, which flows across the feedback resistor. Because one end of the resistor is at ground potential, then the other end is at $V_{out} = -IR$, or

$$V_{out} = -RC\frac{dV_{in}}{dt} \qquad (10.23)$$

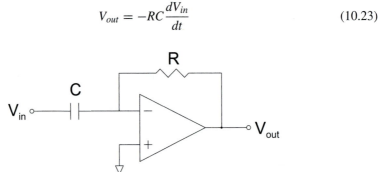

Figure 10.19.   Circuit for using an op-amp as a differentiator.

If you try out this circuit, you will realize the limitations of op-amps. If you apply a 1-kHz triangular waveform to the input with, say, $R = 1\ K\Omega$ and $C = 0.1\ \mu F$, you get a nice square wave at the output, with the low value when the slope is negative and the high value when the slope is positive. The output, however, gets distorted as we increase the frequency. If we replace the triangular wave by a square wave, we have a problem: The slope in a perfect square wave is infinite! The output is, of course, not infinite, as it rises only to the supply voltage. When the output reaches these levels, you may have other issues as well, because the op-amp is out of its comfort zone.

10.2.10 Impedance Transformer

In the next circuit (see Figure 10.20), we inquire about the input impedance. We have a component with an impedance Z connected to the noninverting input of the op-amp.

We define the input impedance as

$$Z_{in} = \frac{V_{in}}{I} \tag{10.24}$$

with I being the current flowing in and through R_1.

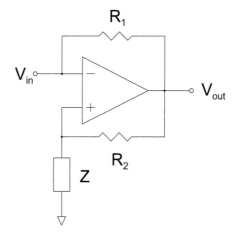

Figure 10.20. Circuit to implement an impedance transformer.

■ **EXERCISE 10.5**

By now, we have all the tools to understand the circuit of Figure 10.20.

1. *Find an expression for V_{in} in terms of Z, R_2, and V_{out}.*

2. *Find an expression for I, the current flowing through R_1, in terms of Z, R_1, R_2, and V_{out}.*

3. *Show that*

$$Z_{in} = -\frac{R_1}{R_2}Z \tag{10.25}$$

If $R_1 = R_2$ and $Z = R$, we get a *negative* resistance: You apply a voltage and the current flows in the opposite direction! Huh? This circuit is also known as a negative impedance converter.

One more trick: If $Z = \frac{-j}{\omega C}$, then $Z_{in} = \frac{jR_1}{R_2\omega C}$; the circuit makes the capacitor look like an inductor. The inductance is

$$L = \left(\frac{R_1}{R_2}\right)\frac{1}{\omega^2 C} \tag{10.26}$$

This is also known as a gyrator. It does not act exactly like an inductor because the final equivalent input impedance depends inversely on the frequency, but we can use a combination of negative impedance converters to generate a true gyrator. They are used in ICs to mimic the effect of an inductor without actually using one.

10.2.11 Complex Feedback and the "Mystery Circuit"

This sounds a bit academic, but the circuit of Figure 10.21 is an interesting one to figure out in the lab (we also have it listed as a lab project). Therefore, we will not spoil the fun, and give you only some clues that help you figure it out.

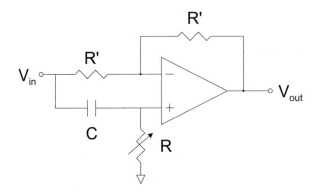

Figure 10.21. This circuit makes an interesting transformation for you to figure out.

■ **EXERCISE 10.6**

Figure out the circuit of Figure 10.21.

1. *Get expressions for V_+ and V_-. Keep all complex numbers in the $a + ib$ form—do not use exponential notation yet.*

2. *Express V_{out} in terms of V_{in}. A bit of algebra is involved here, including complex-number manipulations (nobody said it would be easy). When you get to the final relationship between V_{in} and V_{out}, convert the final complex number to the exponential notation.*

3. *What does the circuit do? (Psst: You get to the answer if you can answer the following two questions: What is the relationship between the amplitudes of V_{in} and V_{out}? What is the relationship between the phases of V_{in} and V_{out}? How does it behave as a function of ω? What are the extreme cases?)*

Hope you were successful. If not, you must do the lab and see it for yourself—there is nothing like the real thing.

10.2.12 Active Filters

In this section, we give an elementary discussion of active filters. Be aware that a huge amount of literature exists about them, and a good design should probably consult a more dedicated treatment. The good news is that we can also make good and simple filters with op-amps. One of the limitations of the filters seen previously, passive filters, is that they have attenuation and high output impedance. Of course, we can fix this with op-amps, because they can provide good impedance matching. In addition, we can amplify and compensate the attenuation that is typical of passive filters. Note that active filters are good at low frequencies (less than 1 MHz). For higher frequencies, we still need to use passive components.

The simplest active filters follow the brute force approach: We put one or more filter stages followed by a noninverting amplifier. Figure 10.22 shows low-pass (left) and high-pass (right) filters in combination with noninverting amplifiers.

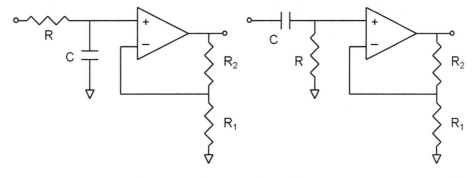

Figure 10.22.   Active filters, low-pass (left) and high-pass (right), with noninverting amplification.

The transmission function is the same as for the passive filters (Equations 7.9 and 7.15), but multiplied by the amplification $1 + \frac{R_2}{R_1}$. The inverting version of both low-pass and high-pass active filters are shown in Figure 10.23.

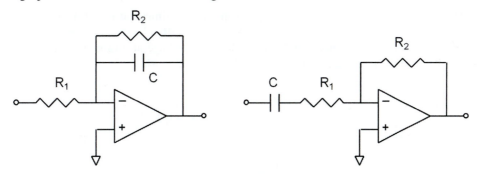

Figure 10.23.   Active filters, low-pass (left) and high-pass (right), with inverting amplification.

For the low-pass filter, the transfer function is given by

$$\frac{V_{out}}{V_{in}} = -\frac{R_2}{R_1(1 + j\omega/\omega_c)}$$

where $\omega_c = \frac{1}{R_2 C}$. You can see that, at high frequencies, $V_{out} = V_- = 0$; at zero frequency, it is $\frac{-R_2}{R_1}$.

■ **EXERCISE 10.7**

Find the transfer function for the inverting high-pass active filter of the previous figure.

A number of designs for second-order filters exist. Figure 10.24 shows the Sallen and Key design. The output is given by

$$V_{out} = \frac{1/(Z_1 Z_2)}{1/(Z_1 Z_2) + (1/Z_1 + 1/Z_2 + 1/Z_3)/Z_4} V_{in} \tag{10.27}$$

For a low-pass filter, components 1 and 2 are resistors (for best performance, set $Z_1 = Z_2 = R$) and components 3 and 4 are capacitors $\left(\text{for best performance, set } Z_3 \sqrt{2} = \frac{Z_4}{\sqrt{2}} = \frac{1}{j\omega C}\right)$.

For a high-pass filter, we switch the components. The cutoff frequencies are given by $\frac{1}{RC}$. For a band-pass filter, components 1 and 3 are resistors, and components 2 and 4 are capacitors. Higher-order filters are obtained by cascading combinations of filters, but such designs are beyond the scope of our discussion.

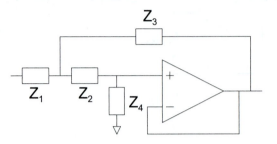

Figure 10.24. Sallen–Key design of two-pole active filter. Depending on which impedances are resistors and capacitors, the circuit is low-pass, high-pass, or band-pass (see text).

10.2.13 Sample and Hold

Often a signal provides information that is piled up at certain times and absent at other times. The signal can vary quite rapidly, so it can be difficult to measure anything. A sample-and-hold circuit helps us "freeze" the input voltage at a time when it is providing the information. The circuit of Figure 10.25 does the task.

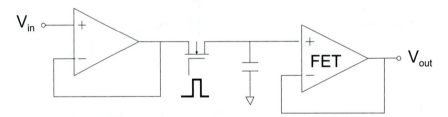

Figure 10.25. Sample-and-hold circuit: At the receipt of a pulse input to the FET, the input voltage is transfered to the intermediate stage charging a capacitor.

When the gate of the MOSFET is enabled, the capacitor is allowed to charge quickly, so $V_{out} = V_{in}$. Then when the gate input goes low, the output of the circuit becomes isolated from the input. Because the capacitor is connected to the noninverting input of an op-amp (with FET inputs), the voltage on the capacitor remains constant. The capacitor will still slowly discharge, because there is a very small current at the inputs of the op-amp.

10.2.14 Voltage Regulators

In Chapter 8 (Diodes), we treated the voltage regulators as black boxes. Now we can understand how they work: They use feedback. The circuit in Figure 10.26 is a simple version of a voltage regulator. A reference voltage sets the final voltage through a transistor driven by a follower. The collector of the transistor and the power of the op-amp are connected to the unregulated input. The extra voltage (ripple and all of that) is absorbed by the voltage drop V_{CE} across the transistor. The op-amp used there is of a special kind that can work near the supply voltage.

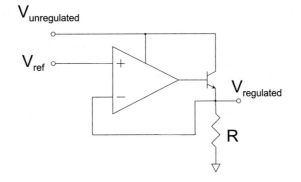

Figure 10.26.   The use of an op-amp to generate a regulated output voltage from an unregulated source.

Switching Power Supply   Another form of regulation with op-amps is implemented for generating power. The circuit of Figure 10.27 is striking for its simplicity and cleverness. An oscillator drives a MOSFET, which grounds an inductor. The inductor produces a voltage spike, but only to give the capacitor a bit of charge through the diode. We have seen this type of trick before with the peak detector. The capacitor permanently discharges through two resistors, forming a voltage divider. The divided voltage gets connected to a follower, which controls the frequency of the oscillator. Thus, the whole circuit stabilizes to a constant (regulated) voltage output. The striking point about the circuit is that the diode somewhat isolates the input so that it can be higher than the unregulated input. This is a simplified version of a switching regulator. The small power converters for cellphones and other consumer electronics are power supplies that operate on this principle, called switching power supplies. They do not care what comes in, including the AC line voltage.

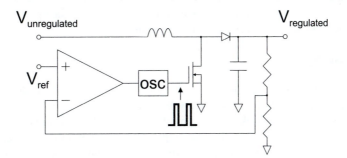

Figure 10.27.   Basic design of a switching power supply.

10.2.15   Feedback Digest

We are ready to move on to other adventures with op-amps, but before we do, let us reflect a bit on feedback. Negative feedback is a means of stabilizing an output, regardless of how we need to accomplish the stabilization. The previous circuit was an example. We live around man-made feedback. Our cars today do many good things for us: cruise control and antilock brakes are a few examples. Without feedback, some modern airplanes and rockets would fly like balloons that we let go of after filling up with air: a quick loop-de-loop and crash. Thus, hugely nonlinear devices can be kept stable with negative feedback. We can make the feedback go beyond the wires. Consider the circuit in Figure 10.28 (see also Lab Project 10.6.4). Instead of connecting the feedback with a wire, we use a light source and detector to make the connection. The interesting aspect of this circuit is that we can "see" the feedback in action. What happens if we move the light source away from the detector? Yes, the light gets brighter. Thus, our feedback can be used to set references other than voltages, such as the illumination of the detector or the speed of the car. Provided that you have the right transducers, this can be anything. The world is out there for us to conquer with feedback.

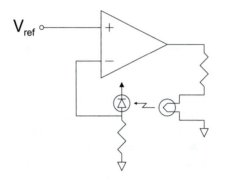

Figure 10.28. We can provide negative feedback by physical means, such as light.

10.3 OPEN-LOOP CIRCUITS

We can do a lot when the negative feedback is partial or nonexistent—that is, when there is a differential input but the resulting output of the op-amp does not stabilize the output. This behavior is predictable, so with some cleverness, we can use the op-amp's response to our advantage. This section gives a few useful examples.

10.3.1 Peak Detector

We saw a version of the peak detector in Chapter 8 (Diodes). The basic design is still the same, but the op-amp cleans things up nicely. In the circuit of Figure 10.29, as V_{in} increases, V_{out} contributes to the feedback by charging the capacitor. If V_{in} decreases, we break the loop because the diode does not allow the output of the op-amp to sink current. The other side of the diode is connected to the noninverting input of a buffer, which draws no current. Therefore, the capacitor stays charged. If we use an FET-based op-amp, the output is stable. Thus, the circuit hangs on at the peak of the input signal.

Because the signal always goes up, we need a way to reset it. To do this, we can put a large resistor across the capacitor so that it discharges slowly. In this case, the output displays the most recent peak of the input. Another possibility is to put a transistor across the capacitor and switch it on when we want to reset the circuit.

10.3.2 Comparator

A comparator is an op-amp working in a full open-loop configuration. We cannot use just any op-amp for this. Op-amps are designed to work closed loop and thus expect a small differential input. If we force the inputs to be different we can damage the op-amp. Certain

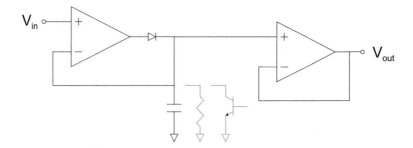

Figure 10.29. Circuit for a peak detector with op-amps

op-amps, called comparators, such as LM306, are specifically designed to work open loop. Comparators can also work with a single supply (but check the data sheets).

If the op-amp has its negative-supply input to ground, the circuit of Figure 10.30 does the following operation:

- If $V_- < V_+$, where $V_+ = \frac{R_2 V_{CC}}{(R_1 + R_2)}$, then $V_{out} = V_{CC}$.
- If $V_- > V_+$, then $V_{out} = 0$.

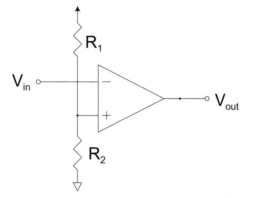

Figure 10.30. Op-amp comparator that compares an input voltage to a positive voltage defined by a voltage divider.

Thus, this circuit can be used to interface with digital. It can be used to make a true/false decision. For example, if we want the output of a circuit to reflect that our input signal is less than some reference voltage, this circuit will do it. We can pick R_1 and R_2 appropriately to get the value needed for the reference voltage.

If we want to detect a transition from low to high or from high to low, this circuit will do that, too. However, if the input signal is noisy, it can produce multiple high/low transitions in the output when the signal fluctuates around the high/low threshold voltage. As we have seen before in digital, we can fix this with a Schmitt trigger. Here is how we can use op-amps to make a Schmitt trigger with distinct high-to-low and low-to-high thresholds that we specify. Consider the circuit of Figure 10.31. When the input is high the output is $V_{out} = 0$.

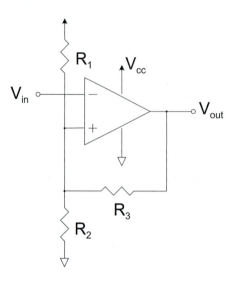

Figure 10.31.   Circuit for a comparator with hysteresis: a Schmitt trigger.

In that case

$$V_{+_{HL}} = \frac{R_2||R_3}{R_1 + (R_2||R_3)} V_{CC} \tag{10.28}$$

By "$R_2||R_3$," we mean the equivalent resistance of resistors R_2 and R_3 connected in parallel. When $V_- < V_+$, then $V_{out} = V_{CC}$. If the input goes low, then the output goes to V_{CC}. Because the output is high, then

$$V_{+_{LH}} = \frac{R_2}{R_2 + (R_1||R_3)} V_{CC} \tag{10.29}$$

Notice that $V_{+_{LH}} > V_{+_{HL}}$; the threshold for going from low to high is larger than the threshold for going from high to low. For example, if $R_1 = R_2 = R_3$, we have $V_{+_{HL}} = \frac{V_{CC}}{3}$ and $V_{+_{LH}} = \frac{2V_{CC}}{3}$. This reduces the chances that noisy transitions will produce multiple pulses.

10.3.3   LM555 Timer

It is now time to see what is inside the LM555. We have waited this long so we can fully understand it. This classic IC oscillator consists of two comparators and a flip-flop. The various inputs can be connected in ways to make this a useful and versatile device. Figure 10.32 shows a block diagram of the inside of the LM555. The circuit has two comparators, C_1 and C_2. Table 10.1 lists the settings of the flip-flop inputs for all combinations of outputs of C_1 and C_2 (always when the voltage on #4 is high).

Knowing how the comparators set the flip-flop lets us understand the circuit of Figure 10.33. It is an astable multivibrator. When $V_2 (= V_6)$ dips below $\frac{V_{CC}}{3}$, pin #7 is not grounded internally, and the capacitor charges through R_A and R_B. As the voltage of the capacitor rises over $\frac{V_{CC}}{3}$, the comparators put the flip-flop in the no-change state so that the capacitor continues to charge. When the voltage on the capacitor goes over $\frac{2V_{CC}}{3}$, the comparators put the flip-flop in the reset state. Its output Q' switches the transistor so that it grounds pin #7. As this happens, the capacitor begins to discharge to ground through

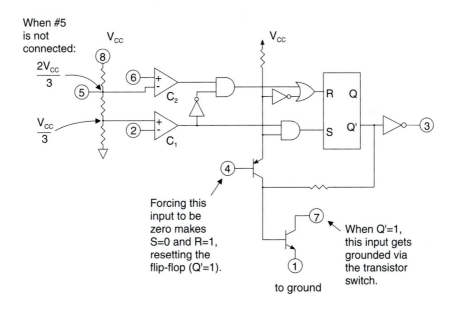

Figure 10.32. Circuit diagram of the LM555 timer IC.

Table 10.1. Operating Truth Table of the LM555 IC

C_1	C_2	S	R	Action
0	0	0	0	No change
0	1	0	1	Reset $(Q' = 1) \rightarrow$ #7 is grounded
1	0	1	0	Set (# 7 floating)
1	1	1	0	Set (# 7 floating)

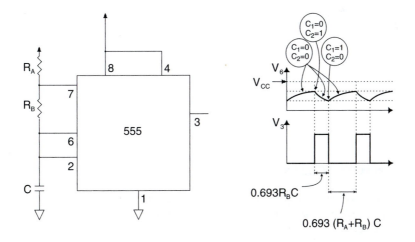

Figure 10.33. Wiring (left) and waveforms (right) of the LM555 operating as a monostable multivibrator.

R_B and through the transistor. When the voltage on the capacitor dips below $\frac{2V_{CC}}{3}$, it puts the flip-flop again in the no-change state. That is, it stays in the reset state with pin #7 grounded. When the capacitor dips below $\frac{V_{CC}}{3}$, we start all over, ungrounding pin #7 and charging the capacitor back again. In summary, the flip-flop regulates the charging and discharging of the capacitor, with respective time constants $(R_A + R_B)C$ and R_BC.

The LM555 can also be used to make a "one shot." This is a circuit that puts out a pulse of variable width for every input pulse. Figure 10.34 shows this circuit. Notice that, in contrast to the previous circuit, when pin #7 gets grounded, it discharges the capacitor immediately. A negative-going input trigger, which is connected to pin #2, gets pin #7 ungrounded, so the capacitor starts to charge through R_A. Nothing changes until the capacitor voltage exceeds $\frac{2V_{CC}}{3}$, at which time pin #7 gets grounded *if pin 2 is back to high*. Thus, the circuit can be used to generate a pulse that is wider than the input trigger. The time constant of the capacitor-charging circuit, $1.1R_AC$, determines the width. Because the input trigger discharges the capacitor, it must have a width that is *shorter* than the desired width of the output pulse. The best example of this application is the automatic door lock. The person inside the apartment pushes a button to unlock the door for a guest, but the door unlocks for only a specified amount of time that allows the guest to get in.

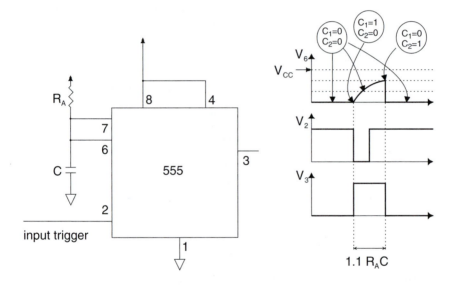

Figure 10.34.   Wiring (left) and waveforms (right) of the LM555 operating as a one-shot.

10.3.4   Relaxation Oscillators

Oscillators use positive feedback. That is, the noninverting input gets to sample the output. As a consequence, the output swings toward the rail. This was done earlier with the Schmitt trigger. We can do better than this by making the output swing back. Consider the circuit of Figure 10.35. If at some initial time $V_+ > V_-$, the output will swing toward $+V_{CC}$. When this occurs, V_+ becomes $\frac{+V_{CC}}{2}$. The capacitor then starts to get charged with a characteristic time constant RC until $V_- > \frac{+V_{CC}}{2}$, at which time the output swings to $-V_{CC}$, making

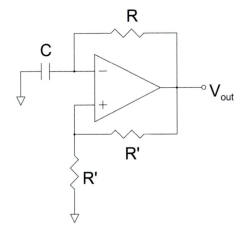

Figure 10.35. Relaxation oscillator.

$V_+ = \frac{-V_{CC}}{2}$. The capacitor discharges until it reaches $V_- = \frac{-V_{CC}}{2}$, when the output flips and it starts charging again. Thus, V_- varies between $\frac{-V_{CC}}{2}$ and $\frac{V_{CC}}{2}$. The output is a square wave that varies between the two rails.

■ **EXERCISE 10.8**

Show that the period of the relaxation oscillator of Figure 10.35 is 2.2RC.

10.4 REAL OP-AMPS

Op-amps are complex circuits. Figure 10.36 shows the diagram of the LM741. It has 22 transistors, 11 resistors, a diode, and a capacitor. This is just one variety of op-amp. Limitations of these manifest in the overall performance of the IC. Be aware of these limitations when designing circuits with op-amps, especially when the circuit is specialized in any way. One of the goals of this section is also to help you understand those apparently incomprehensible electronic data sheets that have all the jargon for strict specification of operating parameters. In this section we discuss specific limitations of real op-amps.

10.4.1 Voltage Gain

The limitation in the gain for a closed-loop circuit is that the closed-loop gain A cannot exceed the open-loop gain G. The frequency dependence of G is important. In low-frequency op-amps, this dependence is made to be predictable, as shown in Figure 10.37 for the LM741 op-amp.

The graph is in $dB = 20\log_{10}\frac{V_{out}}{V_{in}}$, or $\frac{V_{out}}{V_{in}} = 10^{\frac{dB}{20}}$. For example, a gain of 10 is 20 dB, and a gain of 100 is 40 dB. You can see that $G = 10^5$–10^6 for $f = 0$, decreasing rapidly with frequency to about $G = 1$ for $f = 1$ MHz. The graph is in a log-log scale of the gain and frequency, showing that they are inversely related. It reveals the *gain-bandwidth product* (GBP), which is a quantity that is the product of the gain times the frequency (in Hz). This quantity specifies the frequency response of op-amps. The GBP for the 741 op-amp is

schematic

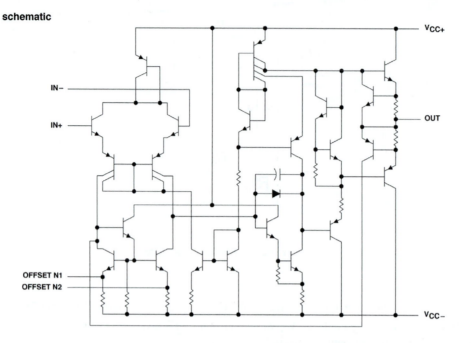

Figure 10.36. Schematics of the LM741 op-amp. Courtesy Texas Instruments.

1 MHz. If we have an application that requires amplification of a signal up to a frequency f_S, then the maximum gain that we can get is:

$$A_{max} = \frac{\text{GBP}}{f_S} \tag{10.30}$$

In the definition of the GBP, f_S is the frequency at which the signal falls by 3 dB, the same definition for the cutoff frequency of filters described earlier. GBP is also known as the unity-gain frequency, or the frequency at which the gain is 1. For example, if we want to build a noninverting amplifier with the LM741 op-amp for signals up to 10 kHz, then the maximum closed-loop gain is $A_{max} = 100$. For LF411, GBP = 3 MHz. Faster op-amps are the LF356 (GBP = 5 MHz) and the LM318 (GBP = 15 MHz).

10.4.2 Slew Rate

This is the maximum speed for rail-voltage output. For the LM741, it is 0.5 V/μs; the LF411 is much better, at 15 V/μs; and a faster op-amp such as the LM318 has a slew rate of 70 V/μs. When the input is a square wave, the op-amp cannot produce an output with an infinite slope. As a consequence, the output looks trapezoidal, with the slope determined by the slew rate. With input sinusoidal waveforms, we get a similar distortion: If the frequency is too high, the output looks more like a triangular wave.

  If we have a sinusoidal voltage $V = V_0 \cos \omega t$, where V_0 is the peak voltage and ω is the angular frequency, the maximum rate of change of the voltage is $\frac{dV}{dt_{max}} = \omega V_0$. Therefore, for sinusoidal voltages, we can define the slew rate as

$$\text{SL} = \omega V_0 = 2\pi f V_0 \tag{10.31}$$

where f is the frequency.

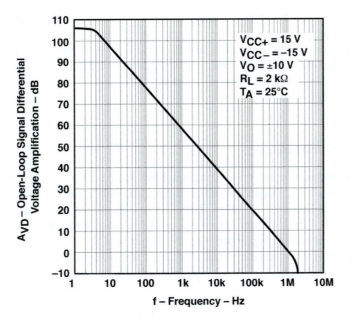

Figure 10.37. Frequency dependence of the gain of the LM741. Courtesy Texas Instruments.

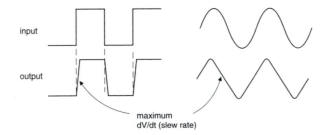

Figure 10.38. Limitation of op-amps for time-varying signals: the slew rate.

10.4.3 Common-Mode Gain

Op-amps are differential amplifiers, which means that, in principle, they amplify a difference in voltage between the inputs regardless of the actual value of these voltages. In practice, this is not so. The amplifier also amplifies the voltage that is common to both inputs, also known as common mode, with a gain G_{cm}. If we define the common-mode voltage as $V_{cm} = \frac{V_+ + V_-}{2}$ and the differential open-loop gain is G_{ol}, then the output of the op-amp is

$$V_{out} = G_{ol}(V_+ - V_-) + G_{cm}V_0 \tag{10.32}$$

Amplifiers are specified by the amount of common-mode rejection that they attain. In op-amps, this is specified by the *common mode rejection ratio* (CMRR), defined as

$$\text{CMRR} = \frac{G_{ol}}{G_{cm}} \tag{10.33}$$

In principle, we want CMRR to be infinite. Standard op-amps have CMRR between 80 and 100 dB near zero frequencies, but lower at higher frequencies. In an inverting amplifier, this is not an issue because the noninverting input is grounded, so the common-mode voltage is zero. In the noninverting amplifier, the input voltage is the common-mode voltage, so here the common-mode gain shows up as an additional amplification. In differential amplifiers, however, the common-mode gain is particularly critical, and the CMRR specification is very important.

10.4.4   Input Impedance

The input impedance for a noninverting amplifier is given by the current entering the noninverting input V_+. The inputs of a BJT-based op-amp go to the base of transistors (see the circuit of Figure 10.36). Thus, it depends on the small but finite base current. For the case of LM741, the input impedance is about 1 MΩ. The op411, which is driven by the gates of FETs, has a much higher impedance of about 10^{12} Ω.

If the amplifier is an inverting amplifier, as shown in Figure 10.39, the impedance is *much smaller*. Because the inverting input is at a virtual ground, the input impedance of the circuit is R_1.

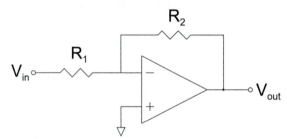

Figure 10.39.   The input impedance of the inverting amplifier is R_1 (in first approximation) because the inverting input is at ground potential.

10.4.5   Output Impedance

The output impedance of op-amps is not low. For the LM741, the output impedance is $R_{out} \sim 75$ Ω. This large output impedance limits the loads that can be driven by an op-amp. In closed-loop circuits, the output impedance can be reduced further depending on the gain of the circuit:

$$r_{out} \sim R_{out}A/G \tag{10.34}$$

10.4.6   Output Current

When an op-amp is set to output a current value that it cannot deliver (connected to a load that is too low), the output voltage drops below the desired value. The output current is $\frac{V_{out}}{R_{load}}$. Figure 10.40 shows the maximum peak output voltage that the op-amp can output for a given load resistance.

We can see that below 1 kΩ, the output voltage sinks to less than 12 V for a supply voltage of 15 V. In addition, when connecting inverting amplifiers in series, we must remember that the input impedance of the circuit is the drop between the input and V_- divided by the input resistor at the inverting input of the op-amp. Thus, if $V_- = 0$, then R_1 must be of the order of 1 kΩ or greater (10 kΩ is best) if we expect high output swings.

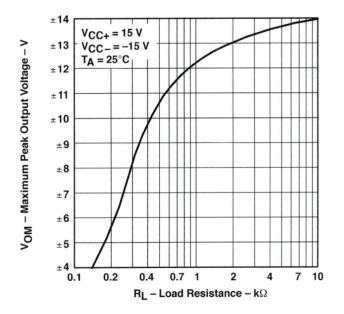

Figure 10.40. Graph of the maximum output voltage as a function of the load resistance for the LM741. Courtesy Texas Instruments.

10.4.7 Input Bias Current

The base of the BJTs and the gate of FETs draw small but finite bias currents. These currents may affect precision circuits. Consider, for example, the inverting amplifier (Figure 10.41). The input current $\frac{V_{in}}{R_1}$ is split into i_- and $\frac{-V_{out}}{R_2}$. The expression for V_{out} is then

$$V_{out} = i_- R_2 - \left(\frac{R_2}{R_1} \right) V_{in} \qquad (10.35)$$

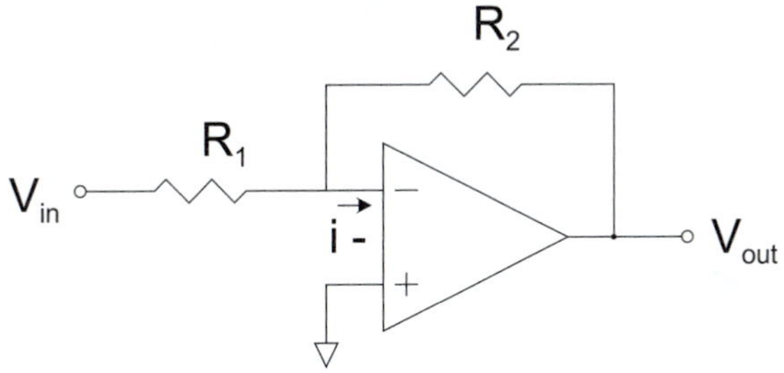

Figure 10.41. Some finite amount of current enters the differential inputs.

If $R_2 = 1\,\text{M}\Omega$, then the effect of the input bias is 0.08 V for the LM741 ($i_- = 80\,\text{nA}$). We cure this by adding a resistor R_3 to the noninverting input and connecting it to ground, as shown in Figure 10.42.

Let us calculate R_3. The currents in the inverting branch of the circuit are

$$\frac{V_{in} - V_-}{R_1} = i_- + \frac{V_- - V_{out}}{R_2} \qquad (10.36)$$

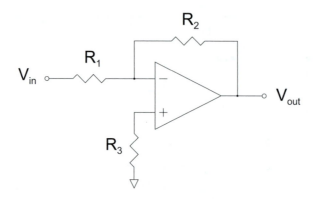

Figure 10.42.   Circuit to correct for the input bias current.

Considering the input i_+ bias, we find

$$V_+ = -i_+ R_3 \tag{10.37}$$

Using $V_- = V_+$ and replacing Equation 10.37 into Equation 10.36, we get

$$\frac{V_{in}}{R_1} + i_+ \frac{R_3}{R_1} = i_- - i_+ \frac{R_3}{R_2} - \frac{V_{out}}{R_2} \tag{10.38}$$

We will recover the noninverting amplifier equation $V_{out} = -\frac{R_2}{R_1} V_{in}$ if

$$i_+ \frac{R_3}{R_1} = i_- - i_+ \frac{R_3}{R2} \tag{10.39}$$

If we assume $i_- = i_+$, then we get from Equation 10.39

$$R_3 = \frac{R_1 R_2}{R_1 + R_2} = R_1 || R_2 \tag{10.40}$$

Op-amps with FET inputs have low input bias currents. For example, for the LF411 it is 50 pA. Some electrometer op-amps have ultra-low input bias currents. The LMC6001 and AD 549 op-amps have input bias currents of 75 fA and 60 fA, respectively.

The input bias currents are not equal, so there is an input offset current. This offset is a fraction of the bias currents. For example, for the LM741, $\Delta i_{+/-} = 75$ pA, whereas for the LF411 it is $\Delta i_{+/-} = 20$ nA.

10.4.8 Input Offset Voltage

If we tie the inputs of the op-amp, we do not get zero output voltage. The output swings to either rail. This is the result of offset input voltages at the input. In precision applications, this problem needs correction.

As we can see in the circuit of the LM741, there are two additional inputs to the circuit labeled "offset N1" and "offset N2." They are connected to pins 1 and 5 of the 8-pin IC package and are used for correcting the input offset voltage.

We correct for the input offset voltage by attaching the ends of a trimpot to the offset inputs of the op-amp, with the center tap connected to the -15 V supply, as shown in Figure 10.43. We adjust the pot until the output is zero when the inputs are tied to ground. We repeat this procedure in lab project 10.6.5.

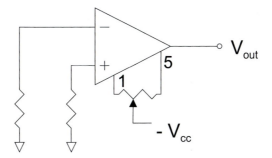

Figure 10.43. Method to eliminate the input offset voltage using the trim inputs of the op-amp.

10.4.9   Power Supply Voltage

Supply voltages do not have a uniform labeling scheme. There are two supply inputs for the op-amp. The positive is normally V_{CC} or V_{CC+} and the negative supply is V_{EE} or V_{CC-}. The usual expectation is that the one supply is $+V_S$ and the other is $-V_S$. Op-amp specifications put limits to V_S. For example, for LF411, $3.5\,\text{V} \leq V_S \leq 18$ V. However, the real limitation is in the difference of the two supplies: a minimum of 7 V and a maximum of 36 V for LF411. We could use supplies with unbalanced voltages ($V_{CC} = +10$ V and $V_{EE} = -5$ V). We could even use a single supply, but you must still keep the inputs within a fraction of the two supplies. Thus, if one supply is grounded, 0 V input is not a good input. A better solution is to redefine your common voltage above zero.

The recent popularity of the single-supply op-amps is convenient for using batteries, but then the common voltage must be switched to something like $\frac{V_{CC}}{2}$. An additional complication of this is that any AC inputs have to be AC-coupled and referenced to the new common. The output has to also take into account the common reference point.

In addition, you must know that op-amps are not designed to go all the way to supply rails. A common specification for this is the *rail-to-rail output*, which specifies how close to the rail the output gets. Manufacturers are not uniform in how they specify this, and often they do not specify the consequences (such as lower speed and gain) of doing so.

Finally, if the power supplies fluctuate, they also introduce a fraction of those changes to the output. The *power supply rejection ratio* (PSRR) specifies the amount that this fluctuation is attenuated. If ΔV_{ps} is the fluctuation in the power supply and ΔV_{out} is the fluctuation that it effects on the output, then

$$\text{PSRR} = \frac{\Delta V_{ps}}{\Delta V_{out}} \tag{10.41}$$

For standard op-amps, PSRR is typically 80–100 dB. It is then good practice to put a bypass capacitor across the supplies to minimize any fluctuation, especially in high-frequency applications.

10.5   PROBLEMS

1. In the circuit of Figure 10.44, find V_{out} in terms of the input voltages.

2. Find the output voltage in the circuit of Figure 10.45.

3. A siren alarm is based on an integrator circuit, shown in Figure 10.46, with a capacitor of $2\,\mu\text{F}$. When the input is a logic value of 1 (5 V), the output of the integrator feeds

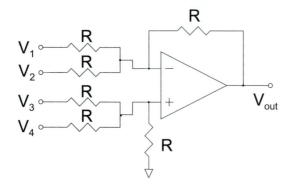

Figure 10.44. Circuit for Problem 1.

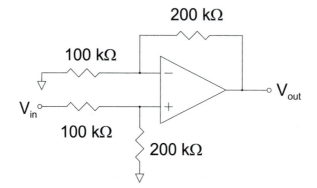

Figure 10.45. Circuit for Problem 2.

onto a siren whose frequency depends on the voltage input to it. Thus, the integrator must:

- Output a voltage that rises linearly with time from 0 to 10 V in 2 s.
- Have a capacitor that must be reset to zero when the integrator voltage reaches 10 V. We do this with a MOSFET switch driven by a comparator.
- You must find:
  - (a) The way to connect the digital input to the integrator so that the output of the integrator rises when the digital input is 5 V.
  - (b) The value of R.
  - (c) The way to wire the comparator. You have only $+5$ V, $+15$ V, and -15 V available.

4. Design a circuit that has $V_{out} = \frac{2}{3} V_{in} + 5$ V.

5. Design a circuit that amplifies the 0.8 V (peak-to-peak) audio signal from an audio circuit that has a 100 kΩ output impedance. The output should be about 10 V (peak-to-peak) and applied to an 8 Ω speaker load.

6. Determine how the circuit of Figure 10.13 works.

   (a) Find an expression for the current flowing through R in terms of V_{in} and R.
   (b) For what value of V_{in} do we have $I = 0$?

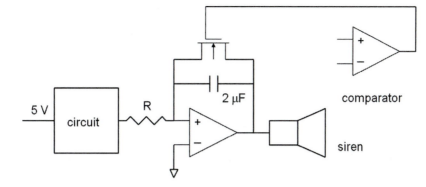

Figure 10.46. Circuit for Problem 3.

7. Draw the complete circuit diagram to generate a reference voltage using the method of Figure 10.15. Use a 3-V Zener diode with a current of 5 mA going through it.

8. A photodiode with a responsivity of 0.9 A/W is to be used to detect light at the 1 mW level. Design a circuit that outputs 3 V/mW.

9. The circuit of page 147 is an instrumentation amplifier.

   (a) Find an expression for V_{out} in terms of V_{out-1} and V_{out-2}. The values of the resistors are given.

   (b) Find expressions for V_{out-1} and V_{out-2} in terms of V_1 and V_2.

   (c) Show that

$$V_{out} = \frac{R_3}{R_2}\left(1 + \frac{2R_1}{R_g}\right)(V_2 - V_1) \qquad (10.42)$$

10. For each of the circuits of Figures 10.47–10.51 explain what is wrong with it and modify it so that it works. Be particularly observant of connections, component values, loading, and any detail that looks odd. Note that the circuit may have more than one problem.

   (a) The circuit of Figure 10.47 is a ×10 amplifier.

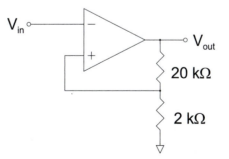

Figure 10.47. Circuit for Problem 10a: a ×10 amplifier. It has one or more errors.

   (b) The circuit of Figure 10.48 outputs −5 V.

   (c) The circuit of Figure 10.49 is using 4-V source to drive a 10-Ω load with 6 V.

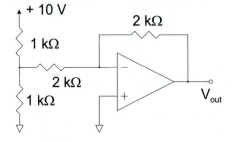

Figure 10.48. Circuit for Problem 10b: outputs −5 V. It has one or more errors.

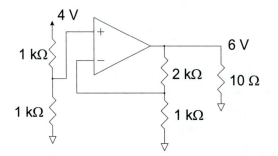

Figure 10.49. Circuit for Problem 10c: drives a 10-Ω load with 6-V generated from a 4-V supply. It has one or more errors.

(d) A ×4 amplifier that suppresses the DC component (Figure 10.50).

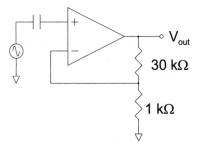

Figure 10.50. Circuit for Problem 10d: a ×4 amplifier that blocks the DC component. It has one or more errors.

(e) Microphone to an 8-Ω speaker circuit (Figure 10.51).

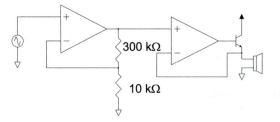

Figure 10.51. Circuit for Problem 10e: driving an 8-Ω speaker from a microphone. It has one or more errors.

11. Explain how the circuit of Figure 10.52 works. *Suggestion*: Analyze the value of V_{out} when $V_{in} > V_+$ and $V_{in} < V_+$.

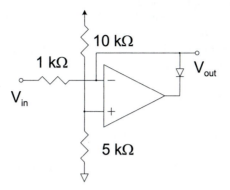

Figure 10.52.   Circuit for Problem 11.

10.6 LAB PROJECTS

This section presents ideas for lab projects.

10.6.1 The Inverting Amplifier

Required equipment: Function generator, oscilloscope, power supply, two speakers.
Required components: LM741 or LF411 op-amp, resistors, 3055 and 2955 power transistors.

1. Set the function generator to a 1-kHz sine wave. Wire an inverting amplifier (shown in Figure 10.53) with $R_1 = 1\,k\Omega$ and with supply voltages of ± 15 V. Set R_2 so that the gain (amplification) A of the amplifier is 1, and verify it by measuring the input and output waveforms with the scope.

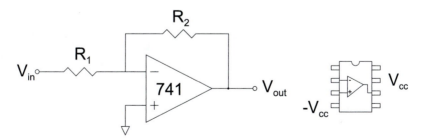

Figure 10.53.   Inverting amplifier with LM741 for lab project.

2. Determine the highest gain that you can achieve. Find this by incrementing the gain by powers of 10 (you may have to adjust the input voltage). Remember that the output voltage cannot exceed the supply voltage.

3. Set the speaker-microphone.

A speaker can work as a microphone. Connect the speaker directly to the scope and observe the waveform that you get. Small amplitudes? Well, all audio equipment must deal with small signals. Send the signal through a suitable inverting amplifier and observe the output. This output is probably ready to be sent to a second speaker (in the correct direction now). However, op-amps do not put out more than 10 mA, so connecting the op-amp directly to an 8-Ω speaker loads it too much. Wire a follower/power booster to drive the speaker.

10.6.2 Noninverting Amplifier

Required equipment: 6-V DC motor with homemade plastic chopper blades, oscilloscope, power supply.
Required components: LM741 or LF411 op-amp, optointerrupter.

1. Figure 10.54 shows an optointerrupter. It has a horizontal base with two pillars rising from the base. An infrared LED is in one pillar and a photodetector is on the other pillar. The LED and photodiode are facing each other, so we use them to determine whether something is in between them. They are normally used to control the revolutions of a motor through a chopper wheel connected to the axis of a motor, as shown in the figure.
   The output of optointerruptor circuits has high output impedances and small output signals. The design given in Figure 10.54 uses $R_1 = 180\,\Omega$ and $R_2 = 220\,\Omega$ for the optointerrupter circuit. Wire it and locate the motor with the improvised chopper wheel, so that it interrupts the path of the light from the LED to the detector. Observe the output of the photodetector with the scope. Connect the motor to a DC power supply and increase the voltage until you see the chopped waveform in the scope (just keep the motor/wheel from becoming a helicopter). *Warning*: Wear safety glasses. It is easy to get the chopper to go to fast and for the chopper blade to crash against the optointerrupter and come out flying.

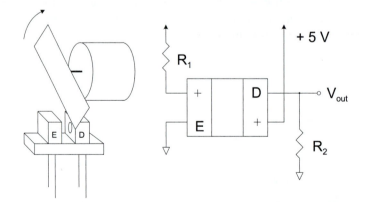

Figure 10.54. Optointerruptor driver circuit for Lab Project 10.6.2.

2. Normally, the revolution control of a motor/chopper is done digitally. Amplify the output of the optointerrupter so that the amplitude of the output waveform is 5 V. Do not disassemble the circuit.

3. Design a circuit that inverts the waveform of the previous section. The inverted output should still be between 0 and 5 V.

10.6.3 Mystery Circuit

Required equipment: Function generator, oscilloscope, power supply.
Required components: LM741 or LF411 op-amp, resistors, capacitors.

Set up the circuit of Figure 10.55 and study the input and output when a 1 V, 10 kHz sine wave is applied to the input. Use a 10-kΩ pot for R and set $C \sim 0.003\,\mu$F. Set R' to 1 kΩ. Your job is to determine the function of the circuit. Do this in conjunction with Exercise 10.6 on page 276.

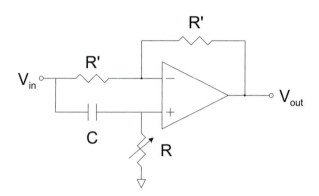

Figure 10.55. Circuit for Lab Project 10.6.3.

10.6.4 Servo and a Constant-Illumination Controller

Required equipment: Power supply.
Required components: LM741 op-amp, BPW77 phototransistor, 6-V filament light-bulb, resistors, capacitors.

Here we design one of the most useful applications of op-amps: a servomechanism. Our design involves controlling a light-detector system, so that the illumination of a light detector by a bulb remains constant. We exercise the use of the op-amp while building the different components. We use six op-amps in this lab, so organize and wire your breadboard neatly.

1. Light Bulb Driver

   We have provided you with a 6 V light bulb that we want to drive with an op-amp. The bulb requires currents that exceed the 20 mA rating of the '741. The circuit shown in Figure 10.56 uses an op-amp in combination with a transistor to drive the light bulb. Complete the wiring so that the light bulb is driven by the input V_1. Complete also the partial schematic in the box of Figure 10.56.

2. Photodetector

   (a) Of the various types of photodetectors, phototransistors are suitable for applications that involve low speeds and high gain. The photo-current generated by the p–n junction of the detector is amplified internally as it becomes the base current

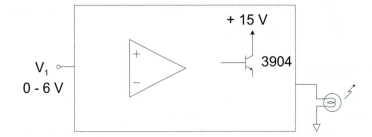

Figure 10.56. Circuit for part 1 of Lab Project 10.6.4.

of a transistor. When biased, as shown in Figure 10.57, the phototransistor acts as a current source. The output voltage V_2 is negative. Why?

(b) Wire the circuit shown in Figure 10.57 so that the photodetector is near the light bulb. Invert the output to generate an output $V_3 = -V_2$, and observe with the scope V_3 as a function of V_1. Complete the diagram.

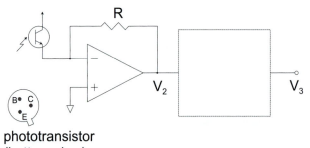

Figure 10.57. Circuit for part 2 of Lab Project 10.6.4.

3. Differential Amplifier

(a) Wire a differential amplifier circuit so that the output $V_4 = V_A - V_B$, where V_A and V_B are input voltages.

(b) Connect the output V_4 to an integrator with a time constant of 1 s, as shown in Figure 10.58.

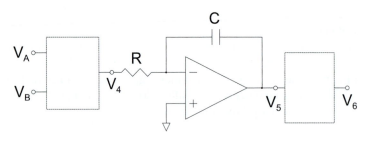

Figure 10.58. Circuit for part 3 of Lab Project 10.6.4.

(c) Invert the output of the integrator so that $V_6 = -V_5 = \frac{1}{RC} \int V_4 dt$.

(d) Connect V_B to ground. Connect a DC power supply to V_A and observe V_A and V_6 with the scope. Explain the response of V_6 when V_A is positive or negative.

(e) Decrease the time constant of the integrator to 1 ms and repeat the previous step.

(f) Disconnect V_B from ground and connect it to V_6. Explain the behavior of V_6 vs. V_A.

(g) Increase the time constant of the integrator to 1 s and repeat the previous step.

4. Putting It All Together
   Assemble the circuit as shown in the block of Figure 10.59. We will complete the feedback loop of this "slow follower" with the light source and detector.

   (a) Set V_A to zero.

   (b) Disconnect V_6 from V_B, and connect V_B to the output of the detector circuit V_3.

   (c) Connect the output of the integrator V_6 to the input of the light bulb circuit V_1.

   (d) Increase V_A until the light bulb starts to glow not too strongly. Now move the light bulb away from the detector. The light bulb should get brighter. Explain why. If you explained it, congratulations—you now understand how feedback allows automation, how the cruise control of your car works, how missiles "lock" to their targets, and much more.

   (e) Reduce the time constant of the integrator and observe the response of the circuit.

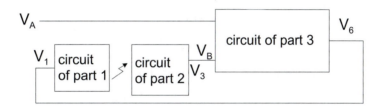

Figure 10.59.   Circuit for part 4 of Lab Project 10.6.4.

10.6.5   Real Op-Amps

Required equipment: Function generator, oscilloscope, power supply.
Required components: LM741 and LF411 op-amp, resistors, trimpot.

1. Gain *vs.* Frequency
   Wire a simple inverting amplifier, as shown in Figure 10.60. Keeping $R_2 = 100\,k\Omega$, choose R_1 for gains of 1, 10, 100, and 1000. Find the 3 dB bandwidth for each gain. Compare your results with the specifications of the LM741.

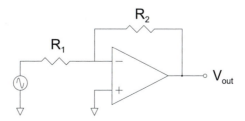

Figure 10.60.   Circuit for part 1 of Lab Project 10.6.5.

2. Slew Rate

   (a) Set an op-amp follower and connect a square wave, with 1 kHz 1 V peak, to its input. Observe the input and the output as you increase the amplitude of the input. Measure the slew rate.

   (b) Reduce the input to 5 V and change the waveform to a sine wave. Observe the output as you increase the frequency. Above a certain frequency, the op-amp will not be able to follow the input. Measure the slew rate from the output above that frequency. Compare your result with the previous one.

3. Input Offset Voltage and Bias Current.

   (a) Wire the circuit shown in Figure 10.61. Measure the output with the DVM. Explain the result.

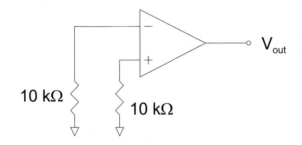

Figure 10.61. Circuit for part 3a of Lab Project 10.6.5.

   (b) Connect a feedback resistor to the previous circuit, as shown, and trim the output so that is zero.

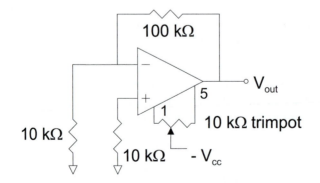

Figure 10.62. Circuit for part 3b of Lab Project 10.6.5.

   (c) Use the feedback resistor and measure the output. Trim it further to have V_{out} as close to zero as possible. Then put back the feedback resistor and measure V_{out}.

10.6.6   Practicum Test

You are given an input AC signal. Use operational amplifiers to complete the tasks listed. Ask the instructor for help if you are stuck. However, be aware that points are taken off for help. Make diagrams of all your wirings.

1. Amplify the AC signal by a factor of three.

2. Add a DC component so that the signal does not go below zero.

3. Use the signal to drive an incandescent bulb. Note that the latter operates at high currents (hundreds of milliamps).

CHAPTER 11

CONNECTING DIGITAL TO ANALOG AND TO THE WORLD

Contents

As a finale for our course, we will consider ways in which we can connect digital and analog. We have already covered a number of applications connecting them: gates driving LEDs and switches driving gates. However, we have not discussed ways to connect digital gates directly to analog circuits. After learning about many aspects of analog circuits and their components, we have all the ingredients to make a final connection to digital.

This last step enables us to connect smart digital devices, such as computers and processors, to analog circuits. We can connect the latter to electro-mechanical devices in the outside world. We must start by understanding all those black boxes that we used earlier in the course: gates. Once we know how they work, we will know how to connect them to analog circuits.

11.1 TTL GATES

In this section we discuss the basic aspects of TTL gates. The direction of current flow is not as obvious as you may think. It also helps us understand the underlying electronics behind the operation of these gates.

The Input Transistor   Much of the inner workings of a TTL gate center on a key component called the input transistor. Figure 11.1 shows a simplified form of a NAND gate.

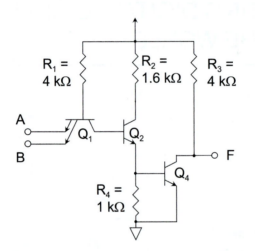

Figure 11.1.   Circuit inside a basic TTL NAND gate.

On the left are two inputs, A and B, that are connected to two emitters of the transistor named Q_1. This is not the most unusual aspect of this transistor. Consider the situation in which A or B (or both) is a logic 0 (that is, with an output ~ 0.2 V). The base-emitter junction of Q_1 gets forward-biased turning on the transistor. Notice that the collector is connected to the base of another transistor. By turning on Q_1, we are drawing (or trying to draw) a current from Q_2. We can indeed do this, but this "reverse-bias" current will be very small. Because $V_{BE1} \sim 0.6$ V, then

$$i_{B1} = \frac{V_{CC} - V_{B1}}{R_1} \sim 1.1 \text{mA} \tag{11.1}$$

However, this means that $i_{B1} > i_{C1}$, which puts Q_1 into saturation. As a result, $V_{CE1} \sim 0.2$ V. Therefore, the base of Q_2 ($V_{B2} = V_A + V_{CE1} \sim 0.4$ V) is not high enough to turn on Q_2. If Q_2 is off, then the base of Q_4 is 0 V (it is connected to ground through R_4), so Q_4 is also off (or in a high impedance state: $R \sim 1 \ M\Omega$), which makes the output F be 5 V by its connection to V_{CC} through the pull-up resistor.

Therefore, if *A or B* is 0, then Q_1 is on, Q_2 and Q_4 are off, and therefore, F is a 1. It is important to realize that when the inputs are low, they *draw* current (~ 1 mA) from the gate.

Suppose now that both are logic 1 (that is, with an output in the range 2.5–5 V). Then the B–E junction of Q_1 is no longer forward-biased. However, the collector-base junction of Q_1 is forward-biased, sending a current *into the collector*, and feeding it to the base of Q_2. If Q_2 is on, then a current flows through R_4. The drop across R_4 turns on Q_4. It can be shown that Q_4 is saturated; therefore $V_{C4} \sim 0.2$ V, or F is a logic 0.

Open-Collector Outputs   In the previous circuit, we could have left out resistor R_3. This is the open-collector configuration mentioned in Chapter 3 (Combinational Logic). A logic 1 in the output is really a high-impedance state because the output transistor is off. The benefit of this is that we can "wire-AND" the outputs of several gates in a place where the outputs of several gates converge, such as in a bus. If an open-collector gate is used alone, we have to provide the pull-up resistor externally to the gate so that a logic 1 is 5 V.

11.1.1 Totem-Pole Output

Although the previous design is a viable one, it is not very fast. The propagation delay, caused by the RC time constant of the output ($R_4 C_{nextgate}$), can be as high as 60 ns. Figure 11.2 shows an improved designed, called a "totem-pole" output.

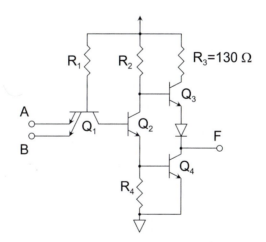

Figure 11.2.   Circuit diagram of a TTL NAND gate with totem-pole output.

Transistor Q_3 improves the switching time to about 8 ns. Here is how it works. When either input is low, Q_2 is off and then the base of Q_3 is 5 V, turning on Q_3 and putting the output at about ~ 3.8 V. When both gate inputs are high, Q_2 is turned on and $V_{B3} = 0.8$ V. Because the output is at 0.2 V, $V_{E3} = 0.8$ V also, so then Q_3 is off. The diode is there precisely to create the drop that prevents the B–E junction of Q_3 from being forward-biased.

11.1.2 Modified Totem-Pole Output

The circuit shown in Figure 11.3 improves the overall capabilities of the previous design. This especially improves the fanout capability of the high output state. The diode is replaced by the transistor Q_5.

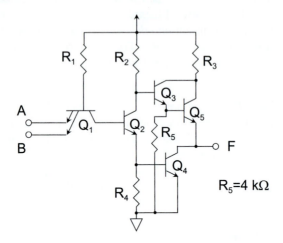

Figure 11.3.   Circuit diagram of a TTL NAND gate with modified totem-pole output.

11.1.3 Tristate Output

Often we want to connect the outputs of gates to the same line, but still keep the fast switching times that are not possible with the open-collector output. The solution is to modify the previous circuit in what is called the tristate configuration, which basically adds an enable to the gate (Figure 11.4). This way, when the enable is asserted, the output is in a high impedance state.

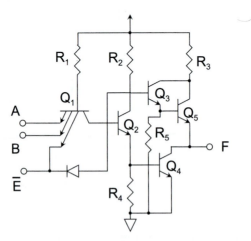

Figure 11.4.   Circuit of the TTL NAND gate with tristate output.

In terms of the previous circuit, when $\overline{E} = 1$, the state of this input does not disturb the gate. When $\overline{E} = 0$, it sets the NAND gate to output a logic 1 (by turning Q_1 on, Q_2 off, and

Q_4 off), but it also turns off Q_3, which keeps Q_5 off. The output is thus isolated. Tristate buffers are the preferred type of gates when connecting devices to the bus of a computer system.

The circuits get more elaborate as technology improves (for example, with TTL-LS and TTL-F), but what we have seen so far is an example of the basic operation of the circuits, which is to make transistors act like switches (that is, turning some transistors on and others off in unison to make the output either a logic 1 or a logic 0). This is even more obvious and simple with CMOS gates, shown next.

Input/Output Characteristics of TTL Gates The input of a TTL gate must *source* current to the input to process an input logic zero. This is because it is connected to the emitter of a transistor. As mentioned earlier, this current is about 1 mA. A circuit connected to the input of the gate that cannot sink current while holding the input voltage low will not be able to drive the gate. When the input is high, the gate draws a small current (a few microamperes).

Transistors drive the outputs of TTL gates, so they can source currents of up to 10 mA. This allows the gate to drive about 10 other gates (but not more!), a capability called "fanout."

11.2 CMOS GATES

CMOS gates use MOSFETs to provide the switching action. The circuits are much simpler because MOSFETs are driven by voltages instead of currents. Consider the circuit for a NOT gate, shown in Figure 11.5. Its simplicity is surprising. It uses two complementary MOSFETs: an n-channel MOSFET (lower) that turns on with a voltage above its source terminal, in combination with a p-channel MOSFET (upper) that turns on with a voltage below its source terminal. This way, a high input (a logic 1) turns the lower n-channel MOSFET on and the upper p-channel MOSFET off, thus connecting the output F to ground (that is, a logic 0). Conversely, when the input is low (a logic 0), it turns the upper MOSFET on and the lower MOSFET off, thus connecting the output F to the supply voltage (that is, a logic 1).

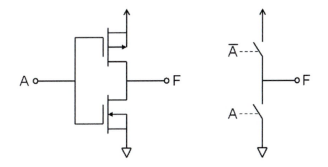

Figure 11.5. CMOS NOT gate.

Consider the next gate: a NAND gate, shown in the circuit of Figure 11.6. When A or B is high, Q_1 and Q_2 are both off, but Q_3 and Q_4 are on, thus connecting F to ground. When either A or B is low, then Q_1 or Q_2 is on, respectively. At the same time that A

or B are low, Q_3 or Q_4 is on, respectively. As a consequence, F is connected to the high voltage and disconnected from ground. The operation can be represented schematically by the input-driven switches shown in the circuit on the right of Figure 11.6.

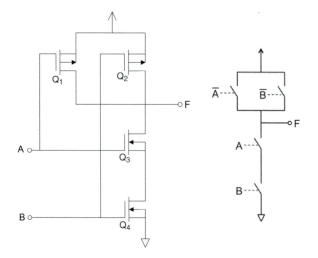

Figure 11.6.   Circuit diagram of a CMOS NAND gate (left), with a conceptual schematic (right).

Input/Output Characteristics of CMOS Gates   Input CMOS gates are connected to the gates of MOSFETs, which are voltage driven. As a consequence, they do not draw or source any current. CMOS inputs have diodes connected to the supply voltage and ground to clamp any voltage above and below the supply limits. This is to protect the gates from voltages built up by static charges. The output of the CMOS can source and sink sizable currents of tens of milliamperes, but at the cost of creating a resistive drop across the output MOSFET.

11.3   INTERFACING

We can couple digital and analog in two fundamentally different ways. Digital I/O is the on/off type that we just discussed: A logic output turns something on or off, and, conversely, a logic input reveals a yes/no type of input. The other way involves analog-to-digital (A/D) and digital-to-analog (D/A) conversion. In A/D conversion an analog voltage is converted to a set of digital lines representing a number proportional to the input voltage, whereas in D/A conversion a set of input digital lines representing a number are converted to a voltage. The input or output analog voltages can in turn be connected to the real world via transducers.

11.3.1   Analog Driving Digital

From the previous discussion on TTL gates, we know that when we apply a logic 1 to a gate, we need to source a very small current (0.05 mA), and when we apply a logic 0, we have to be prepared to sink about 1 mA. Obviously, a logic 0 results in more power being consumed. Therefore, if we want to perform some task and consume the least power, the off state must be a logic 1 rather than a logic 0. This is the reasoning behind low-active inputs and outputs. The keys of a keyboard, for example, are switches that are wired optimally, as shown in Figure 11.7. Because CMOS gates do not draw any current, we could also put the resistor on the ground side of the switch shown in Figure 11.7.

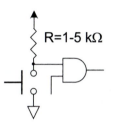

Figure 11.7.   TTL gate driven by a push-button switch.

If we want to drive a gate from an op-amp, we can use the circuit of Figure 11.8.

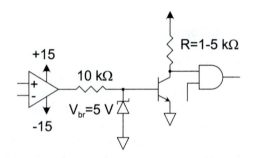

Figure 11.8.   Circuit to drive a TTL gate with an op-amp.

In this case, the transistor prevents the gate from being driven below zero, and the Zener diode keeps the base from being above 5 V. Finally, if we use the circuit to switch devices on and off periodically, using a Schmitt-trigger gate is a good idea.

11.3.2   Digital Driving Analog

Because the TTL gate is a better sink than source of current, it is more convenient to connect the gate to an analog device in such a way that a logic 0 turns on the device. Figure 11.9 shows an example.

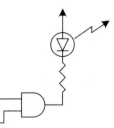

Figure 11.9.   Circuit for TTL gates driving an LED.

If the load requires a current greater than 50 mA, it is better to drive a transistor switch, such as the one shown in Figure 11.10. From our discussion of CMOS, it may seem that because they use MOSFETs, we could drive large currents. This is not true. All components have their power ratings, so we should not use CMOS gates to drive low-impedance loads.

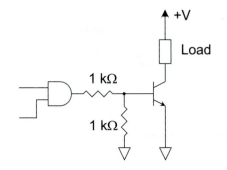

Figure 11.10.   Circuit for a digital gate driving a low impedance load.

11.3.3   Analog-to-Digital Conversion

Another way to interface digital to the real world is to actually drive or sense real devices via analog voltages and transducers. However, in this case, we do not want to interact with the transducer in an on/off mode; we want to control or sense it at a number of levels. For example, if we are sensing heat, we may not want to know just whether something is hot or cold; we might want to know the actual temperature. The converse would be to control the heat delivered to a system at various levels instead of turning the heater on and off. A thermistor, for example, can be used to convert temperature into a voltage. Conversely, the temperature could be controlled by varying the power delivered to a heater. Both situations require us to be able to convert analog voltages to binary numbers, and vice versa. D/A and A/D converters perform these tasks. A/D converters change analog voltages to n-bit numbers. That is, setting the logic level of n digital lines. The converse is true with D/A converters. Both types of converters are illustrated in Figure 11.11.

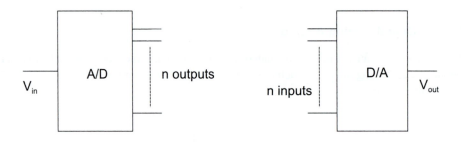

Figure 11.11.   Schematic of electrical connections in analog-to-digital (left) and digital-to-analog (right).

Two important parameters specify the general operation of A/D and D/A. These are *resolution* and *sampling rate*.

Resolution   If the maximum voltage that an A/D can take is V_{max}, and the digital output consists of n bits, then the minimum voltage increment V_{step} or resolution is given by

$$V_{step} = \frac{V_{max}}{2^n - 1} \tag{11.2}$$

That is, it determines what voltage change will result in a unit change in the binary number. This is an important consideration when choosing a converter, because any input voltage variations less than V_{step} do not produce any change in the output. Table 11.1 gives the percent resolution for different bits, with $R = 100 \frac{V_{step}}{V_{max}}$. The same is true with D/A conversion.

Table 11.1.   Resolution of the D/A Converter as a Function of Number of Bits

n	$R(\%)$
2	33.3
3	14.3
4	6.7
6	1.6
8	0.39
10	0.098
12	0.024

Figure 11.12 shows an example of an input waveform that is digitized with a 3-bit A/D. The top graph shows the input analog signal. The middle graph shows how the input voltage is digitized: by changing the number whenever the digital threshold is crossed. The third graph shows the digitized output. Notice that the result is not optimal because the discrete sampling misses some features (between 1 and 2 ms, and between 7 and 9.5 ms). If we want to see the smooth bump between 7 and 9.5 ms, we need to have a step voltage that is less than the size of the bump, or $V_{step} < 1.07$ V.

■ **EXERCISE 11.1**

What is the step voltage in the previous example?

Sampling Rate   Nyquist theorem states that, to reproduce a wave, we need to sample it at least twice every period. Said differently, the sampling rate has to be higher than twice the highest frequency component of the wave. A/D and D/A have different conversion times t_c, so when choosing a converter, we need to ensure that

$$t_c < \frac{1}{2 f_{max}} \tag{11.3}$$

If we do not carefully select the sampling, we could encounter a situation like the one shown in Figure 11.13, where the input waveform can be quite distorted with a sampling rate that is too low.

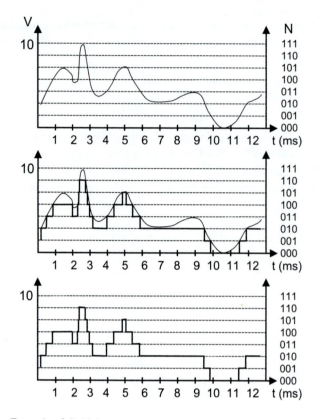

Figure 11.12.   Example of digitizing an analog signal with a 3-bit analog-to-digital converter.

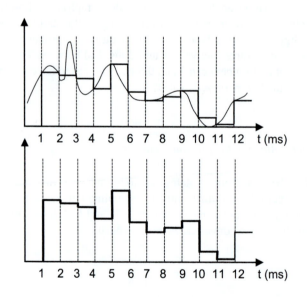

Figure 11.13.   Example of the effects of undersampling a signal.

■ EXERCISE 11.2

What is the sampling rate in the previous example?

You can see that the low sampling rate of the previous example completely misses the sharp peak between 2 and 3 ms. The waveform has a complex shape. Its Fourier spectrum consists of many frequencies, something we discussed in Chapter 7 (Filters and the Frequency Domain). However, we can crudely say that the sharp peak is caused by a frequency component equal to the reciprocal of twice its width, or $f = \frac{1}{(2w)}$. For example, if $w = 0.5$ ms, then $f = 1$ kHz. If we apply the Nyquist theorem, we conclude that the sampling rate must be at least 2 kHz, or every 0.5 ms or less. In Figure 11.14, we use 4-bit resolution and a sampling rate of 0.5 ms. You can see that the reproduction of the signal is much better.

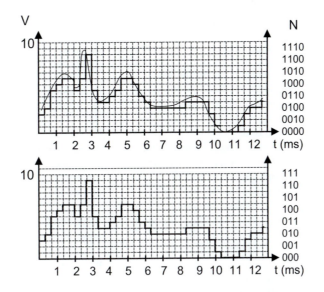

Figure 11.14. Example of a signal that is sampled and digitized adequately.

Undersampling of a periodic wave can also give rise to a waveform that has a period much lower than the original waveform. This is called aliasing. For example, consider the case when a wave of frequency f is sampled with a frequency lower than f, as shown in Figure 11.15. The result of undersampling is a wave with a frequency that has nothing to do with the original signal.

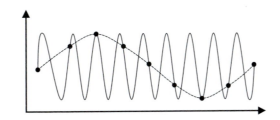

Figure 11.15. Example of aliasing produced by severe undersampling.

11.4   INTERFACING THE WORLD

As digital is taking over much of the electronic devices today, we need to consider the ways in which digital signals are coupled with the real world. This means that we need to convert digital information to and from an analog equivalent. We saw the different forms of accomplishing this. The remaining step is also important: *transducers*. These devices connect the real world, physical quantities, with analog electronic signals. Table 11.2 gives a few examples of transducers. Some transducers are passive (that is, they do not need external power to function), but many others are active. For example, some velocity sensors police use to detect cars' speed emit electromagnetic waves (microwaves), and measure the Doppler shift in the reflected waves. They convert the resulting signal into a digital form that is displayed for the officer to read.

Table 11.2.   Physical Forms and Their Transducers

Physical Form	Sensor	Actuator
Temperature	Thermocouple, thermistor	Thermoelectric cooler, ohmic heater
Motion, position, speed, acceleration	Inductive sensor, capacitive sensor, optical encoder, microwave, radar, lidar, ultrasonic, switches	Piezo electric, magnetic, electromagnetic, motor
Light, microwaves, and radiowaves	Photoresistors, photovoltaics, photodiodes, photomultipliers, ionization, scintillation, thermopiles, bolometers, pyroelectric, antennas	Incandescent bulbs, light emitting diodes (LED), lasers, gas discharges, liquid crystals, electro-optics, antennas
Magnetism	Compass, Hall effect, inductive, resonance	Magnets, electromagnets, coils, wires
Audio	Microphones, speakers	Speakers, electro-mechanics
Pressure, strain	Capacitive, conductive, resistive, optical	Pumps, hydraulics
Chemical	pH sensors, catalytic, conductive, optical, microbalance	Electrolysis, spark igniters

■ **EXERCISE 11.3**

List all the possible sensing transducers in a car.

In many cases, transducers generate analog signals that are processed to directly drive actuators. We have seen many of those in this course (for example, the darkness sensor of Section 1.11.1). In these cases, there is no need to convert signals to digital. However, when some "thinking" is needed, the output of the transducers must be converted to digital and input to a processor or computer. We have avoided a discussion of the interface modules that actually do this because it is a subject in itself, involving both hardware and software, and because the available technologies differ considerably.

■ **EXERCISE 11.4**

Describe a system controlled by a computer that uses transducers. Describe also the role of the computer in relation to the signal received or sent to the transducer.

Finally, transducers have their own characteristics. Sensors generate either currents or voltages, each with their own specifications, including output impedance, so their output may not be ready for direct conversion. Equivalently, actuators have their own input impedance and expect a certain kind of signal with its own specifications. The point is that we must know the specifications of the sensors and actuators and then build analog circuits that interface them with the conversion circuitry. But that we know how to do: It has been the topic of this course.

11.5 PROBLEMS

1. Suppose that both the inputs of the plain TTL NAND circuit of Figure 11.3 are logic 1s.

   (a) As we mentioned earlier, a consequence is that transistor Q_1 is off, but current goes into the collector from the base of Q_1. As a result, Q_2 is on and Q_4 is on. Although we have not demonstrated it, let us assume this is correct. Find the potential of the base of Q_2. (*Hint*: What is the voltage drop at p–n junctions?)

   (b) The base of Q_2 receives current that comes from the base of Q_1. Thus, the B–C junction of Q_1 is forward-biased. What is the base current on Q_1?

   (c) If Q_2 is in saturation, what is the collector current of Q_2? (*Hint*: What is V_{CE} in a transistor that is saturated?)

   (d) How much current flows through R_4?

   (e) Now let us assume that the base current of Q_2 is the base current of Q_1 (part (b)) plus 0.1 mA. What is the base current of Q_4?

   (f) If Q_4 is in saturation, how much current flows into Q_4 from the collector when the output of the gate is not connected to anything?

   (g) Fanout is the capability of a gate to drive other gates. It is typically the number of gates it can drive. In this case, driving means sinking 1.1 mA per gate (see Equation 11.1.). If the transistor Q_4 has $\beta = 25$, what is a reasonable fanout, keeping in mind that we need to keep Q_4 saturated? (*Hint*: What is the saturation condition?)

2. Find the function of the CMOS gate shown in Figure 11.16.

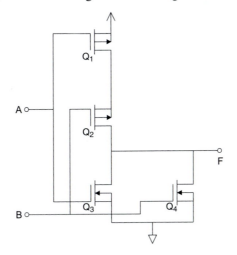

Figure 11.16.   Circuit diagram of CMOS gate for Problem 2.

3. A motor with an impedance of 25 Ω must be powered with the 5 V from a (1) digital TTL gate or (2) CMOS gate. One input to the motor should be 5 V, and the other motor input must be at ground potential. Draw the diagram that is needed to energize one of the motors. Justify the components that you use.

4. For the cases shown, explain in detail which of the circuits is better than the other.

   (a) TTL digital input I/O:

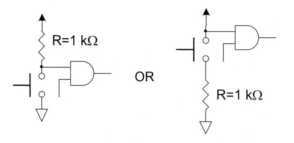

   OR

   (b) TTL digital output I/O:

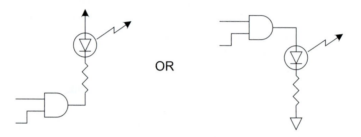

   OR

5. Your current employer hires you to design the alarm system of an ultra secure vault to store secret disks sought by none other than Ethan Hunt (Mission Impossible). The alarm system has two parts: a sensor unit and logic unit. Your employer gives you the specifications outlined. However, Dr. Evil is blackmailing you to provide him with a back-door entry that he will give to Ethan Hunt as part of another blackmail scheme. The specifications are the following:

   (a) <u>Sensors</u>. Four sensors must produce a logic 1 (5 V) when: (1) weight on the floor inside the vault on a special slab detects a person's weight, (2) the temperature of the vault exceeds some low value (that is, detects a person's body temperature), (3) the sound level inside the sound-proof vault exceeds a minimum (pin-drop sound) level, and (4) the path of a laser beam reflected by an elaborate set of mirrors and ending on a light detector is broken. The sensors work the following way:

   - The **weight sensor** is a switch that closes when a person steps on the floor slab of the vault.

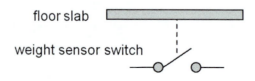

- The **temperature** sensor is a resistor (thermistor) whose resistance value decreases as the temperature increases. The trip point for body temperature is 10 kΩ.

thermistor   1k Ω

- The **sound sensor** is a microphone that puts out an AC signal. The trip sound level has an amplitude of 50 mV (peak).

microphone

- The **light sensor** is a phototransistor that, when illuminated by the laser, puts the transistor into saturation.

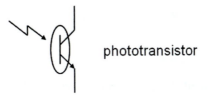

phototransistor

You must devise a circuit for each of these sensors. That is, each circuit takes the input in the form specified and translates it to a digital logic value of 1 (5 V) when the trip point is exceeded (for temperature and sound) or the condition is met (person on the floor or laser-beam path interrupted).

(b) The logic of the alarm system must include:

- If the weight sensor circuit puts a 1, then the alarm gets turned on.
- If two or more of the other sensors put a 1, then the alarm gets turned on. This is to avoid fluke trips produced by a bug.
- For the back-door entry, if all the sensors get triggered at the same time, the alarm does not get turned on. That is, if Ethan steps on the floor of the vault and turns on a hair dryer, flips on a boom box, and blocks the laser beam simultaneously, he is safe to get the disks.
- The output of this logic circuit must switch a 10-V 70-Ω relay that turns on the alarm system.

Design a circuit that implements the alarm system. It has the sensor logic outputs as inputs and must drive the relay to turn on the alarm system.

6. If we have a digitizing scope that has a 10 V, 6-bit, 10 ns conversion time A/D at its input, what are the minimum requirements on the input signal such that the scope is able to measure it?

7. Suppose that we are designing a computer-automated temperature controller.

   (a) The temperature gauge gives an output voltage given by $V_g = 0.1T$, where T is the temperature in degrees Celsius. For temperatures between 0 and 100 °C, a 1–10 V A/D can be used. What is the minimum number of bits the A/D converter must have to be sensitive to temperature variations of 0.2 °C? Justify your answer.

   (b) A digital circuit compares the digitized number to a number that represents a set temperature. If the reading is below this temperature, the comparator outputs a logic 1; otherwise it outputs a logic 0. Design a circuit that will drive a 15 V 1 A heater when this happens. Assume that the digital output comes from a typical TTL gate. Draw the circuit diagram and explain its operation.

8. In a modern type of telephone, the signal generated by the speaker is filtered such that audio above the highest human voice's frequency of 4 kHz is suppressed. After the filter, there is an amplifier followed by an 6-bit A/D converter.

   (a) What is the maximum conversion time of the A/D?

   (b) If the maximum voltage for the A/D is 10 V, what is the minimum detectable voltage change in the amplified audio signal.

11.6 LAB PROJECTS

This section presents ideas for lab projects.

11.6.1 Stepper Motor

Required equipment: Unipolar stepper motor.
Required components TTL gates, power transistors.

Stepper motors differ from CW motors in that they move in a controlled way. They have four sets of coils. In powering one set of coils of the motor, the rotor is locked in a fixed position. When power switches to a different set of coils, the motor locks to a different angle. This way, the motor position can be stepped predictably and reliably. Printers, computer disc drives, and many other devices use stepper motors to move components in a precise way.

The stepper motor that we have is a unipolar motor, which has four sets of coils. It is stepped forward by powering each coil consecutively. Table 11.3 shows the forward sequence. Each step involves a rotation of 7.5 degrees. However, to energize the coils, we need to pass 200 mA through them. This is done by switching a voltage of 5 V to the proper coil.

Design a digital circuit that drives the stepper motor:

1. Design a push-button switch that generates digital pulses.

2. Design a circuit with four low-active outputs that get consecutively active with each cycle of the pulse generator.

3. Use 3055 npn transistors as power boosters for powering the coils of the stepper motor from the outputs of the digital circuit.

4. Put it all together.

5. Replace the output of the push-button switch with a LM555-based pulse generator.

Table 11.3.   Sequence of Steps to Energize a Unipolar Stepper Motor

Cycle	A1(Orange)	A2(Blue)	B1(Yellow)	B2(Black)
1	0	1	1	1
2	1	0	1	1
3	1	1	0	1
4	1	1	1	0

[Red = ground]

11.6.2   Connecting to the Analog World

Required equipment: Function generator with VCO input, speaker.
Required components: AD558, LED, 180 Ω resistor, TMM2114 memory, DP8216 bus driver.

Procedure   This lab has the same layout of the lab in Section 5.12.2. Refer to it for more information.

Digital I/O   Connect an LED to a 4-bit counter in a way that shows that the counter is counting. Connect an SPST push-button switch to an LED so that it controls the LED indicator such that, at each push of the button, it toggles between being enabled and being disabled.

D/A

1. Wire the circuit to the D/A chip (AD558) with pins 9, 10, 12, and 13 wired to ground; pin 11 wired to power (5 V); and pins 14, 15, and 16 wired together. The last three will be the output of the D/A.

2. Connect the four least significant bits of the D/A to ground.

3. Connect the input to the four most significant bits of the D/A to switches of the logic board.

4. With a voltmeter, record the voltages that the D/A puts out for given digital inputs. Take several readings. From this, deduce the minimum step voltage.

Audio Playback   Here we make some electronic music. Refer to the block diagram in Figure 11.17. Dashed lines are connections to be made at different times.

1. Connect the output of the D/A to the voltage-controlled oscillator input of the function generator. The voltage applied will control the frequency of the function generator.

2. Connect the output of the generator (1 kHz, sinusoidal) to the speaker.

3. Set up a 74LS190 decade counter to count up freely and connect the outputs of the counter to the four most significant inputs of the D/A. You should now hear some musical tones played by the counter. Try it out. If you want all 16 states, replace the 74LS190 with the 74LS191. The latter is a binary counter with the same pin connections as the 74LS190.

4. This part involves a circuit used previously (Section 5.12.2), where you store numbers to the memory IC TMM2114. Now disconnect the D/A from the counter. Connect

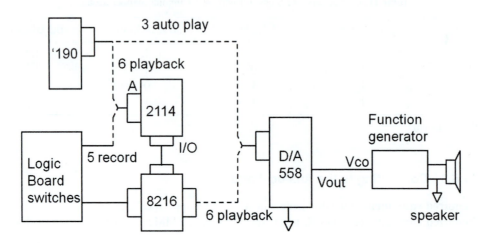

Figure 11.17.   Block diagram of the circuit for lab project 11.6.2.

the outputs of the DP8216 bus driver (also connected to the TMM2114) to the inputs of the most significant bits of the D/A. This way, we can have the numbers stored in memory drive the audio tone. Try it.

5. Now store a set of numbers (encoding a song?) in the lowest ten memory locations of the memory IC (keep the other address lines A4-A9 low). Do this by setting the memory to "read" by advancing the memory locations by hand via switches in the logic board.

6. Now it's playback time: Disconnect the address inputs of the memory from the switches and connect them to the outputs of the counter. The speaker should now be playing your recording—digital music!

APPENDIX A

LOGIC BOARD

The logic board is a basic working tool for doing digital electronics labs. It consists of a 5-volt power supply, a set of 16 debounced switches that connect the outputs of the supply (0 V or $+5$ V) to a 16-pin socket, and a set of 32 buffered LEDs that are driven by $+5$ V board inputs.

The circuits for each switch and LED are standard. The double-pole-double-throw switches are debounced by an octal debouncer IC MAX6818. The inputs to the LEDs are buffered through a standard TTL hex inverter (74LS04).

The logic board is inexpensive, at \$50 (without the cost of the power supply), but laborious to build. Still, the job is nothing that an in-house technician cannot handle. For the power supply, we recommend a 5-amp (minimum) supply, with overcurrent protection. Figure A.1 shows our own homemade version made in the 1980s at Colgate University. The debouncer ICs used in that board are the DM8544 switch debouncers, which are now obsolete. Commercial versions are available for a few hundred dollars.

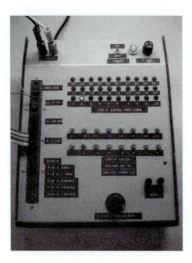

Figure A.1. Photo of a homemade logic board.

APPENDIX B

IF THE CIRCUIT DOES NOT WORK

Contents

If your circuit does not work, do not worry: It happens to everyone. Most of the time the circuit does not work because of a trivial reason, so go through the following checklist.

B.1 DESIGN

Do you have a circuit diagram? It is easy to get carried away and just plug things in. You must have a diagram. Have you checked analytically that the circuit should work?

If you are unsure of this, you better get this part straight. Otherwise, you may be wasting valuable time.

B.2 THE OBVIOUS

Check power and ground to all chips. Follow all the wires. Misplacing a wire in a breadboard is easy. Also check that the power switch is turned on. Some analog circuits require a positive and a negative supply. The plus and the minus of a single supply will not do. Some supplies have a separate ground connection. That connection is *not* connected to the plus or minus outputs of the supply. If you use several supplies, make sure that all the common lines (signal ground) are connected together.

B.3 PLACEMENT

For digital circuits, check that the chips are positioned correctly. An indication that you wired a chip incorrectly is that the chip is very hot (but beware of burning yourself, because sometimes they come close to being red hot). If you do find a chip in this condition, it may not be good anymore. Put in a new chip and throw the old chip in the trash.

B.4 PINS

Make sure that the pin connections are what you think they are. Some ICs have nonstandard connections. Transistors have no uniform pin assignments.

B.5 BREADBOARDS

Sometimes the breadboards get a bit worn out and the contacts are no longer good. Some breadboards may look like a combat zone. You know you have breadboard problems if the output changes when you jiggle the chip in its socket. If this happens, wire the chip in a different part of the breadboard or, if possible, get a new breadboard altogether. Best practice for digital circuits is to use a *new* breadboard at the beginning of the course.

B.5.1 Past the Obvious

If you have a sophisticated circuit, you must now go to the circuit diagram. Follow the wiring using the labeled circuit as an aid. By now, a small subtlety is causing the damage. There *is* a reason for everything. Sloppy and messy wiring can lead you to make mistakes. Chase the problem by using a logic probe or the oscilloscope.

B.5.2 Digital Circuits

A low-cost logic probe for digital circuits consists of a long wire connected to an LED indicator in the logic board. This way, it lights up if the logic level of the line is a 1. If beyond this you also connect the wire to an inverter followed by another LED indicator (preferably of a different color), then you have a light that will turn on with each logic level. Knowing both states of the line is helpful because sometimes a bad connection or no connection is indicated by both lights being off or on.

Test the circuit by putting the probe after a gate and verifying that its output has the expected logic value when you change the state of the input variables. If it is not what you expect, put the probe at the inputs and check that those have the values you expect. Eventually, you will find the source of the problem: a bad connection, an IC that is not wired correctly, two outputs connected together, or a poorly powered IC. Experience will make you an expert in this troubleshooting procedure.

B.5.3 Analog Circuits

An oscilloscope is essential to debugging analog circuits. You must look at the waveform or voltage level at different parts of the circuit. Be careful where you connect the ground of your oscilloscope. It is connected to Earth ground. If one of the supplies is connected to ground also, you should not use the grounding of the oscilloscope in any part of the circuit that is not at ground potential. Otherwise, the oscilloscope will change the circuit.

Some complex circuits with feedback are tricky, in that different parts of the circuit can affect each other. Separate the circuit into independent modules and test those modules separately. Then add them to the larger circuit (that works) one by one.

A sign of trouble is that transistors are hot. Be careful with electrolytic capacitors hooked up backward. Those capacitors will not work properly and will blow up at some point.

B.6 ABUSIVE POWER

If your circuit is *massive*, you have other issues. Is the power supply capable of driving all the gates? Check to see if it is supplying close to the maximum current. If so, you may have to use more than one supply. This is especially true when you are driving LEDs or seven-segment displays. Transistors can also pass high currents. You can solve this by powering different parts of the circuit with the separate supplies and *connecting the grounds of all the supplies together*.

B.7 STUCK

If all else fails, try replacing ICs and components, but this is usually wishful thinking. It may be a good idea to get some help. The instructor may be able to bail you out, but then he or she will spoil the fun.

B.8 DONE!

If you find the problem, you have some cleaning up to do. A good professional cleans up his or her mess. First, try the circuit: It is easy to fix one thing and mess up something else, so try the full operation of the circuit. Second, discard any bad components. Putting them aside and forgetting about them is too easy (good ones and bad ones look alike): Then they end up back on the shelf, ready to nail some new innocent victim (including you, again). Third, modify your circuit diagram with the change (you do have a diagram, right?). Now you can celebrate and take pictures of your solved problem!

APPENDIX C

CURVE TRACER

Contents

A fundamental way to understand nonlinear electronic components is via their I–V curve, so we need to measure it. The ideal instrument is a *curve tracer*. However, it is an expensive instrument. An inexpensive solution is to do individual measurements. In this appendix we present a procedure for getting the I–V curve of diodes and transistors.

C.1 I–V CURVES FOR DIODES

Diodes have a characteristic I–V curve, consisting of a near-zero curve below the forward voltage drop V_d, and a near-vertical curve at V_d. For standard diodes, $V_d = 0.6$ V, but for LEDs, this varies between 1.5 V and 4 V. The curve for Zener diodes is the same in the forward direction; in the reverse direction, it is near zero until the Zener voltage V_Z, where it is nearly vertical. The left side of Figure C.1 shows a simple circuit for measuring the I–V curve of diodes. It involves a DC power supply, a resistor, a voltmeter, and an ammeter. The latter two measure the diode voltage V and current I, respectively. Alternatively, we

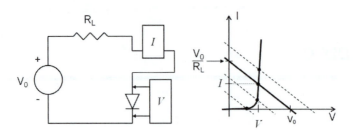

Figure C.1.   Circuit (left) to measure the I–V curve of a diode (right).

can measure the voltage across the resistor V_R and use $I = V_R/R_L$. The resistance R_L is fixed, with a convenient value of $R_L = 220\ \Omega$, and the DC voltage V_0 is variable. This way, by varying V_0, the load line translates (see the graph on the right of Figure C.1). The intersection point of the load line and the diode curve is the measured voltage across the diode and the curve flowing through it. This method is more forgiving on the power supply than the technique of applying the voltage directly to the diode.

This curve can also be measured with the oscilloscope by replacing the DC power supply with an *ungrounded* 60-Hz AC source. We then use the two channels of the scope with one channel (1 or X) measuring the voltage across the diode and the other channel (2 or Y) measuring the voltage across the resistor (which is proportional to the current). The ground of the scope *has to be* the point between the diode and the resistor (see Figure C.2). This is because all scopes have their grounds tied together. Then we set the scope to measure in XY mode. Because of the grounding constraint the current will appear to be going negative. Most scopes have an invert function (for channel 1/X) that can be activated to plot the curve properly. The photo of Figure C.2 is of a 7.5-V Zener diode. For AC source, we used a 12.6 V tube-filament supply.

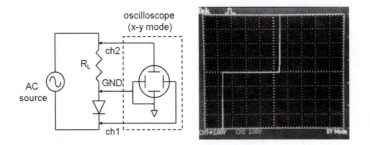

Figure C.2.   Circuit diagram for a curve tracer of a diode using an oscilloscope (left). Photo (right) of the I–V curve of a 7.5-V Zener diode taken with the circuit.

You can see clearly the two voltage drops of the diode: the forward one at ~0.8 V and the reverse drop at the 7.5-V Zener voltage.

C.2   I–V CURVES FOR TRANSISTORS

The transistor needs two circuits, one to send a constant current into the base and another one to bias the collector. The idea is to do something similar to the circuit for obtaining

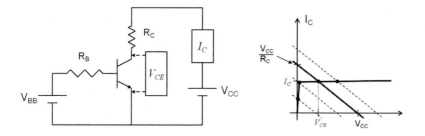

Figure C.3. Diagram of the circuit (left) to take data to graph the I–V curve (right) of an npn transistor.

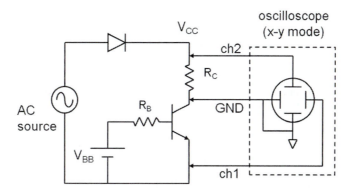

Figure C.4. Circuit diagram of the transistor curve tracer with an oscilloscope.

the I–V curve of the diode. The bias for driving the base is a separate circuit. Figure C.3 shows the specific method for npn transistors.

Because $I_B = (V_{BB} - 0.6)/R_B$, if we make $V_{BB} = 5$ V, then if $R_B = 440\,\text{k}\Omega$, $I_B = 10\,\mu\text{A}$. If we make $R_C = 2\,\text{k}\Omega$, then by scanning V_{CC}, we get to measure (I_C, V_{CE}) pairs. For example, if we assume $\beta = 100$, then transistor action occurs at 1 mA. As we scan V_{CC}, the saturation region will be below 2 V. We can crudely estimate that these parameters scale with the base current. We can replace R_B by a 0–500 kΩ variable resistor and thus adjust I_B. We can also measure I_B by either measuring it with an ammeter or measuring the voltage across R_B.

Getting the transistor curve in the oscilloscope is tricky but doable. We need two floating supplies, one AC and the other one DC. Figure C.4 shows the circuit. The AC source supplies a variable V_{CC} that is kept positive by a diode serving as a half-wave rectifier. The DC supply provides a constant base current, which can be adjusted by either changing the V_{BB} and keeping R_B fixed or the converse.

A key point is that the collector of the transistor is at ground potential, specified by the oscilloscope (see Figure C.4). This arrangement graphs one transistor curve. The photo on the left side of Figure C.5 shows the actual wiring of the circuit, which as you can see, is quite simple. The photo on the right shows a sample curve obtained with the components mentioned previously. You can read the scale of the oscilloscope: Horizontal is 0.2 V/division; and vertical is 1 V/division, or equivalently, 0.5 mA/division using the value of the resistor on the collector (2 kΩ).

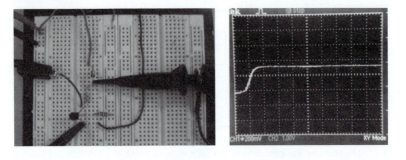

Figure C.5.   Photos showing the actual wiring of the curve tracer for the transistor (left) and the I–V curve for a 3904 transistor (right).

Multiple curves are not necessary because with a variable supply for the base, we get the desired curve by just adjusting it. However, this is still doable. The objective is then to supply different voltages V_B. We can do this with a more elaborate circuit:

- Generate a trigger pulse from the AC source. An easy way is by diode-clipping V_{CC}; we can use a 5-V Zener so that the input to the next stage does not exceed 5 V (see Section 8.5.6).

- Send the pulse to a 2-bit binary counter. We can do this with a standard counter or with two JK flip-flops (see Section 5.6.4).

- Send the 2-bit output of the counter to a 2-bit D/A converter. Again there are other cheap alternatives, such as sending the MSB and LSB to an op-amp summing amplifier (Section 10.2.4), where the output will be $-R_F(V_{MSB}/R_{MSB} + V_{LSB}/R_{LSB})$. This is the basis of a standard R-2R ladder digital-to-analog conversion. In this case, $R_{LSB} = 2R_{MSB}$. So if $R_{MSB} = 5\,\text{k}\Omega$ and $R_F = 3\,\text{k}\Omega$, then the output voltages will be 0, -1.5 V, -3.0 V, and -4.5 V. Because these are negative, we need to invert them. The circuit has the added complexity of having to power the op-amp. More practical is to get a standard D/A powered by 5 V.

- The output of the D/A is connected to the base of the transistor through a resistor. This way, the transistor bias current is cycled every 60th of a second, with the sequence repeating every 15th of a second.

This makes for a nice project!

FETs can also be used either by removing the resistor at the base and varying the base voltage, or by keeping a constant base voltage and feeding it to a voltage divider that outputs to the FET gate. For pnp BJT transistors and p-channel FETs, we reverse the polarity of the supplies and the diode in the collector/drain circuit.

INDEX